The
BERRY GARDEN

The BERRY GARDEN

Cultivation, Decoration and Recipes

TEXT BY MARY FORSELL
PHOTOGRAPHS BY TONY CENICOLA

FOREWORD BY ROSEMARY VEREY

Macdonald Orbis

© Running Heads Incorporated 1989

First published in Great Britain in 1989
by Macdonald & Co. (Publishers) Ltd.
London & Sydney

A member of Maxwell Pergamon Publishing Corporation plc

THE BERRY GARDEN
was conceived and produced by
Running Heads Incorporated
42 East 23rd Street
New York, New York 10010

Editor: Sarah Kirshner
Designer: Liz Trovato
Production Manager: Michelle Hauser

British Library Cataloguing in Publication Data

Forsell, Mary
The berry garden.
1. Great Britain. Berries
I. Title   II. Cenicola, Tony
582'.0464

ISBN 0-356-17975-3

Macdonald & Co. (Publishers) Ltd
Headway House
66–73 Shoe Lane
London EC4P 4AB

Typeset by Trufont Typographers.
Colour separations by Hong Kong Scanner Craft Company, Ltd.
Printed and bound in Singapore by Times Publishing Group.

0 9 8 7 6 5 4 3 2 1

The descriptions and photographs of berries provided in this book are not intended to
be used for purposes of the identification of berries in the wild, and this book is also
not intended to encourage novices to pick their own berries in the wild unless those
berries are to be used solely for ornamentation. Information on how to identify
different kinds of edible berries is covered in comprehensive region-specific field
guides, a variety of which are listed in the Bibliography. This book is also not intended
to advocate the medicinal use of berries or to offer medical advice. Information related
to these topics is provided as historical background only and no one should attempt to
ingest any substance for its purported medical benefit without seeking competent
medical advice and/or diagnosis.

ACKNOWLEDGMENTS

Many talented and knowledgeable people contributed generously to this book and deserve far more detailed acknowledgment than this space allows me to mention.

At Running Heads, editor Sarah Kirshner brought a fresh perspective to the book and provided many welcome suggestions. Marta Hallett and Ellen Milionis stirred up enthusiasm for the project—from start to finish. Ellen Watson was crucial to the research for the book. Michelle Hauser contributed her production skills and culinary advice. And Kate Struby tirelessly tracked down missing information. Liz Trovato's gorgeous design speaks for itself. Timely advice from writer-editor Anne Halpin helped set the book on course.

Bantam editors Coleen O'Shea and Fran McCullough patiently provided their expertise. Thanks also go to Becky Cabaza for her unfailing good cheer and capable work.

Other individuals contributed immeasurably for diverse reasons. They are writer Naomi Black; researcher Sue Manby; Harry Baker, fruit officer of Wisley Garden; Dan Sealy of Cape Cod National Seashore/Province Lands Visitor Center; Dr. Nicholai Vorsa of the Blueberry and Cranberry Research Center; berry grower Alexander Illitch Eppler; berry grower G. R. Roberts of New Zealand; consultant Arthur Nersesian; the Burke, Burke & Burke law firm; Edmond Moulin, director of horticulture at the Brooklyn Botanic Garden; Nancy Morris, site manager of Van Cortlandt Manor; Sue Sanders of Cornell Plantations; Dean Norton of Mount Vernon; Linda Gandrup of Country Essences, Watsonville, California; Susan Leibowitz; Paul Stojanovich; Seemo Moisio and the Finnish Food Institute; Bill Merchant, the Bass Shop; Mark Sonzogni; Joanne Singley; Francine Lerner; and Florence Katz.

Numerous American berry growers allowed us to photograph their beautiful farms and gardens. They are as follows:

California: Nita Gizdich, Gizdich Ranch, Watsonville; Joan and Mike Clifford of Family Ranch, Watsonville; David Dixon, Dixon Ranch, Graton, California; Perry Kozlowski, Kozlowski Berry Farm, Forestville.

New York State: the Greig Farm, Red Hook; Lee Reich, Ulster County; Tantillo's Market, New Paltz; and Rick and Franka Bishop, Cook's Falls.

Thanks also go to Britain's National Trust and National Garden Scheme for allowing entry to Barnsley House, Upton House, Tintinhull House, and East Lambrook Manor Garden. Additional thanks to the Forde Abbey. We are also indebted to private homeowners, including Lady Hobhouse; Mr. and Mrs. Simister of the Harford Lodge; Mr. and Mrs. Seel of Great Rollright Manor; Mrs. E. Hope of Ibstock Close; Mrs. Anthony Sanford of Poulton Manor; the Lord and Lady Vestey of Stowell Park; Sir Charles and Lady Mander of Little Barrow; and Mrs. Shepherd of Great Rollright. Thanks also to gardeners Neil Hewertson and Roger Chapman.

For their recipes and hospitality, we thank the following restaurants, inns, and individuals:
Centerbrook, Connecticut: Steve Wilkinson, Fine Bouche.

Cape Cod, Massachusetts: Miya Patrick, the Charles Hinckley House, Barnstable; Glen Martin and Dana Browne, the Red Inn, Provincetown; Andrea Johnson, the Bramble Inn, Brewster; and Bob McMaster and Joan DiPersio, the Green Briar Jam Kitchen, East Sandwich.

California: Susan Bashel and Julia Orenstein, Pacific Heights Bar and Grill, San Francisco; Tom Grego and Elena Salsedo-Steele, Ventana Inn, Big Sur; Brenda and Tom Hearn, Belle de Jour Inn, Healdsburg; John Ash and Spencer Dolan, John Ash & Co., Santa Rosa; Jim Gibbons, Jacob Horner, Healdsburg; Michael Wild and Missy Wood, BayWolf, Oakland; Michael McClernon and Andrea Sarto, Paloma, Berkeley; Bill Huneke, Rio Grill, Carmel; Oouida Dorr, San Anselmo.

New York City: Robert Berkley; Christer Larsson, Aquavit; Alfred Portale, Gotham Bar & Grill; Deborah Aker, Sarabeth's Kitchen; and Mark Lahm and Franklin Tweedie, Henry's End Restaurant (Brooklyn).

And in England, we gratefully acknowledge the culinary and gardening advice of Patricia Davies-Gilbert.

Many floral designers also freely gave of their time and talent. They are:

New York State: Ellen Leon, New York City; Patricia Reppert and Vonnie Bragg, Shale Hill Farm, Saugerties; and Jarita's Florist, Woodstock.

California: Susan Baldwin, Aromas; Susan Ragsdale, Gonzales Floral and Gift Company, and Ellen Pelikan, Pelikan Spring Farm, Sebastopol.

Cape Cod, Massachusetts: Jackie McInnis, the Little Country Store, East Sandwich.

Thanks also to Aphrodisia in New York City for allowing us to photograph their wares. We're also grateful to the Monterey Vineyard, Gonzales, California, for permitting us to photograph on their grounds.

CONTENTS

FOREWORD

I was sitting at my desk one day last autumn when Mary Forsell looked in, asking if Tony Cenicola might photograph the berries in the garden. My first reaction, that there was no more summer fruit in the vegetable garden, was quickly put to rights, as Mary led me outside to the red *Rubus phoenicolasius* growing through our pond railings. From the moment she described her ideas, I was intrigued, and they have set me thinking about the way in which we treat berries: what a great diversity there is and how much they can contribute, whether it be to the garden or to the larder.

During summer the edible varieties will be ripening in the *potager*, before the ornamental ones take over in the autumn borders. Strawberries, raspberries, and gooseberries—the latter grown as standards at Barnsley—offer themselves for picking, and it is easy to neglect the beauty of their bright fruits in anticipation of their delicious taste; homegrown and fresh, they have twice the flavor. All the labor of planting and weeding is forgotten as the bushes yield their harvest, and the diversity of color, texture, and shape in the berries we choose makes growing them enjoyable as well as worthwhile.

Just as bulbs are the first sign of spring, so the ornamental berries—on hollies, roses, viburnums, pyracanthas, sumacs and snowberries—mark the approach of winter, reflecting the wild hips and blackberries appearing in woods beyond the garden. They are as essential to the fall as the turning leaves—and more welcome to the birds who flock to devour them. At Barnsley we have a *Rosa rugosa* hedge whose generous hips create a second season of interest just as important as its deep pink summer flowers. The tall *R.* 'Wickwar', too, is dramatic when its tawny hips stand out against the sky. Later on, *Malus toringoides* will have a haze of yellow berries covering its bare branches.

This coming year I will be noticing even more, for my eyes have certainly been opened to the wide range of both edible and ornamental berries, some cultivated and some wild in the fields and forests. Although the height of the berry season is midsummer, Mary Forsell shows us how the season can be extended well into the autumn and even into the coldest months of winter through careful selection, preserving, and freezing.

This is a book-of-all-trades, encouraging you to look at berries afresh and to be more adventurous in your choices. It also gives practical hints, recipes, and flower arrangements, and I much enjoyed the historical review of the tradition of berry growing, stretching back as far as 5000 B.C. Whether you consider yourself first and foremost a gardener, a cook, or a photographer, here is a wealth of interest and information.

Rosemary Verey
Barnsley Close
Gloucestershire, England

Chapter One

THE TRADITION OF BERRY GROWING

EUROPEAN ORIGINS

*E*ven before berries were cultivated in gardens, they grew plentifully in the wild. Paleobotanists have discovered that Mesolithic peoples in 5,000 B.C. residing in what is now Poland, Germany, Scandinavia, and England enjoyed a variety of berries in their diet and ate them as a supplement to their hunting spoils. Later, in the Neolithic period in Europe—around 3,000 B.C.—people supplemented their diets, which consisted primarily of meat and grains, by gathering raspberries, blackberries, bilberries, strawberries, rosehips, and elderberries. In Siberia, prehistoric peoples stored berries in icy pits to ensure a fruit supply during winter, an ancient form of "putting food by."

By eating berries fresh from the wild, these early peoples were deriving maximum healthful benefits from them. Berries are very nutritious and contain valuable minerals and carbohydrates. Calcium, magnesium, potassium, iron, and phosphorus are among the minerals contained in most well-known berries. They are also valuable sources of vitamins. Black currants, rosehips, and strawberries are rich in vitamin C. Gooseberries, blackberries, and blueberries are an excellent source of vitamin A. Berries are also low in calories, with blueberries being on the high end with seventeen per ounce and gooseberries low at eleven per ounce.

A confusing term, *berry* really has two meanings, one popular, the other scientific. In the strictest sense, a berry is a fleshy or pulpy fruit that usually does not split open and has one or more seeds but no stone (pit). It develops from a single enlarged ovary in a plant. If we adhered to this definition, strawberries, blackberries, and raspberries would be considered impostors because they are the products of several carpels belonging to a single flower. Grapes, tomatoes, and eggplant, for example, could claim to be berries along with currants, blueberries or bilberries, and cranberries as well.

For our purposes, we will not quibble about seeds and stones and will regard as a berry almost any small fruit whose popular status as a berry is unquestioned, irrespective of its structure. This eliminates from our discussion such fruits as cherries and grapes—subjects best left to books on orchards and viticulture—but includes unusual though traditionally recognized berry fruits like rosehips, Cape gooseberry, rowanberry, and Japanese Wineberry.

After civilizations had formed and the most rudimentary horticultural practices had been established, wild berries were still ardently consumed—but not cultivated. In Denmark, Iron Age peoples gathered wild cranberries for drinks. Egyptian tombs have yielded funerary offerings of berries—no doubt considered an essential in the afterlife—but no evidence has been found of their having been grown in the home garden.

The exception to the rule is mulberry cultivation. Ovid, who died in the year 17, alludes to a mulberry tree growing in a

European black currants have been in cultivation for centuries. This lush planting in front of a British stone farmhouse, left, thrives in the cool, moist summers characteristic of England.

Rosehips, above, were among the berries that ancient peoples consumed ardently as much as 5,000 years ago. The high vitamin C content in rosehips bolstered the neolithic northerners' health.

garden in his *Metamorphoses*. In the letters of Pliny the Younger, who died in the year 113, the Roman statesman boasted that the garden at his country house, referred to as *villa rustica*, was thickly planted with mulberry trees.

Although an early form of the kitchen garden—where food, flowers, and fruit are grown—came into being in England during the Roman occupation, there is no evidence that berries other than the mulberry, which the Romans brought with them, were cultivated. Sixth-century Saxon calendar illustrations for the month of September show what may very well be mulberries being harvested from the abundant kitchen garden.

For centuries, berries were regarded as being utilitarian or medicinal. Only the elite enjoyed them for taste alone. While Henry III of England (reigned 1216–72) often enjoyed mulberry and raspberry drinks, these berries—as well as wild elderberry, bilberry, and blackberry—were commonly used as coloring agents for pottages (meat-filled stews), pigment for paintings and manuscripts, or purgatives and laxatives. Berries were not considered foods in their own right.

Edward I of England (reigned 1272–1307) was the first on record to transplant strawberries and raspberries from the wild into the manorial garden for cultivation. Partly as a result, the strawberry was one of the first fruits to graduate from medicinal status to food. It slowly became a fashionable treat among the nobility and later among the middle classes. By the fourteenth century, distinctions had come to be made between herb gardens, orchards, and "noble" gardens, which by that time contained gooseberries and most likely currants. Although elderberries, blackberries, and bilberries never really gained a foothold in these gardens because they were so plentiful in the wild, the berry was beginning to make significant strides as a cultivated fruit.

The Elizabethan period, in particular, provided a congenial atmosphere for the berry to prosper in the British manorial garden. Estates began to develop extensive walled kitchen gardens to satisfy the increasingly sophisticated gastronomic leanings of the gentry—for delicate berry puddings and salads that were usually brightened by strawberry leaves.

One estate, Nonsuch Palace in Surrey, developed an enormous kitchen garden that was famous in its day. The berries in cultivation were strawberries, raspberries (known then as *hindberries*), gooseberries, and barberries. Intricate—and often completely erroneous—horticultural theories developed in England surrounding berries. Gooseberries and mulberries were commonly grafted onto other fruit-producing plants; it was believed that the presence of these berries would influence the host fruits, causing them to ripen early.

The well-heeled sixteenth-century British garden commonly contained barberries, mulberries, raspberries, currants, strawberries, and whortleberries (bilberries). But these were in the gardens of the well-to-do; the cottager was more likely to get by with less variety. The trend toward growing berries had been initiated, and berries slowly made their appearance in smaller plots. The didactic poet Thomas Tusser documented this period in his ambitious work of 1557, *Hundreth Good Pointes of Husbandrie*. Therein he instructs the cottager wife to plant strawberries, gooseberries, and barberries.

Outside of England, meanwhile, the mulberry was the favored berry in the stately gardens. A Venetian ambassador visiting Belvedere Court at the Vatican in 1523 noted the presence of mulberries in the garden. Italian Renaissance villa gardens concealed utilitarian spaces, so it is unknown if berries were grown in the kitchen gardens that were hidden by hedges since no visitors to the gardens were able to document the contents. The situation was not much different in France in the sixteenth century; grand estates such as Gaillon and Blois boasted hundreds of stately mulberry trees on their grounds, but

Kitchen gardens have been the traditional location for berry growing since at least the fourteenth century. This classic walled garden at a private home in England includes raspberries, currants, and blackberries, which are trained along the wall.

bramble and bush berries are not recorded features in the garden.

By the early seventeenth century, berries in the garden had become so commonplace in England that Francis Bacon, in his essay "Of Gardens," was able to envision a special wild garden where the ground is "set with Violets, Strawberries, and Primroses" as well as with wild plantings of red currants, gooseberries, and bearberries—"here and there, because of the Smell of their Blossom. . . ."

Late in the century, the strawberry industry became a major commercial endeavor in England. The berries were sold in London markets, such as Covent Garden and Farringdon market, by women who migrated to the area for the strawberry season. These women—who came from Shropshire on the Welsh border, from Wales itself or even from Ireland—came to be known as "Shropshire girls." They would leave the farms on the outskirts of London at dawn with baskets weighing as much as forty pounds on their heads and walk to London at a brisk pace in groups of twenty or more, arriving in London in time for morning sales, crying, "Strawberries, scarlet strawberries!"

By the eighteenth century, berries were widespread in gardens elsewhere in Europe. In his autobiography, Goethe recalls the vine-trellised garden of his childhood in Germany. Lining the walls was "an interminable row of currant and gooseberry bushes" that successively produced harvests until autumn, as well as an "old, high, wide-spreading mulberry-tree." Later, we find Count Leo Tolstoy reminiscing about his childhood garden in Russia, which contained raspberries. Until the eighteenth century, black currants, which grow wild in France, were gathered from the wild for medicinal purposes and the leaves made into herb teas. The berries were crushed by priests and physicians alike to cure a variety of illnesses, such as fever and plague. But by the nineteenth century, they had become an important wine crop, particularly in the Burgundy region.

During the nineteenth century a quick succession of developments caused berries to evolve substantially. The Victorian craze for hybridizing focused new attention on the berry, particularly that old favorite, the strawberry, and new varieties of berries were developed at an astounding rate in Europe and America. Hybridizing work inevitably spread to the *Rubus* genus of bramble plants to tame some of the thorny traits and make them more garden-friendly. The publication of William Robin-son's *The Wild Garden* in 1870 also encouraged allowing natural-looking plantings of bramble berries—previously not thought of as landscaping elements—into the garden. Robinson's contemporary, the garden writer/designer Gertrude Jekyll, advocated planting bilberries and strawberries as ground cover and enthused about such familiar yet often overlooked plants as the elder, in writings spanning from the late nineteenth through the early twentieth century.

Evidence of currant cultivation in British gardens extends as far back as Elizabethan times, and it was no doubt practiced even earlier. This tart berry, opposite page, was later introduced to French, German, and Russian kitchen gardens.

The first American colonists were baffled by the New World's broad selection of wild berries, which grew everywhere from low, swampy areas to clifftops, left. Native Americans helped them distinguish between edible and potentially toxic species, such as those pictured.

AMERICAN INNOVATIONS

Colonial pioneers experienced an initial period in the New World that they called "starving time." Despite America's bountiful game, fish, and plant life, the first settlers—most of them city born and bred—were not prepared to survive in an untamed land. They were also unfamiliar with many of the plants in the New World.

There were some berry plants that the Europeans did recognize, however, such as the strawberry and raspberry. They were somewhat acquainted with huckleberries and blueberries because of their similarity to the European bilberry. Captain John Smith, the seventeenth-century English colonist and explorer, took an interest in the flora of the New World, documenting currants, mulberries, gooseberries, strawberries, and barberries growing wild. Another early botanical observer was Roger Williams, who dissented with the Massachusetts Puritan establishment and founded Providence, Rhode Island. Williams was particularly taken with the strawberries of North America, which he had not seen before in the wild. He wrote that strawberries were the "wonder of all fruits growing naturally in these parts. One of the chiefest doctors of England was wont to say that God could have made, but God never did make, a better berry. In some part where the natives have planted them, I have many times seen as many as would fill a good ship within a few miles compass."

But there were many more berries—such as sumac and cranberry—that the colonists were unfamiliar with. They were also limited in their knowledge of how to take full advantage of those berries that they recognized from their homeland. The colonists needed to adapt to survive, and they also needed guidance, which came from the Native American.

The Native American taught the settlers how to hunt as well as how to forage for food, prepare, and preserve it. The natives also instructed the colonists in how to concoct natural medicines. Indians used the bounty of natural berries in innovative ways. First and foremost, they ate berries fresh and cooked with them. The Iroquois so prized berries, particularly the huckleberry, that they ventured into poisonous-snake-infested areas to procure them, but would first smear their moccasins with hog fat, which was supposed to frighten away the vipers. Indians mixed strawberries with meal and made strawberry bread. They also concocted numerous cranberry dishes, many involving nuts.

Another popular Indian usage of fresh berries was to create flavored "waters" from them. Rather than imbibing them for pure enjoyment, Indians thought of these "waters" as strengthening tonics. Perhaps the most useful culinary trick that the Indians taught the colonists was how to dry berries. Native Americans commonly dried berries over a fire or in the sun, which preserved them for winter use. They added the dried berries, especially blueberries and barberries, to their breads, soups, puddings, and pemmicans (dried mixtures of meat and fat) much in the same way we add sun-dried tomatoes and raisins to dishes today to impart taste and color.

The Native Americans utilized every part of the berry plant—and not just for consumption. They created pigments from berries, often mixing blueberries with nuts to create a rich brown dye. They introduced the colonists to barberry leaf tea and salad. The Iroquois dried and steeped elder flowers as a tisane (a tea of flowers), boiled the plant's inner bark for use as a painkiller, and rubbed themselves with the elder's dried crushed leaves as an insect repellent. Indians also chewed raspberry bark to combat dysentery. The natives harvested the berries of *Ribes cereum*, squaw currant, to treat stomachaches and other common ailments and boiled the twigs and leaves for varied medicinal uses.

Indian uses of berries had a profound effect on the way the colonists lived. Not only were they able to take advantage of the berries in cooking year-round, but they also adopted many of the Indians' medicinal uses and passed them down through the generations. For example, Catawbas made a medicinal tea from sumac berries, and the Potawatomis used the leaves as a sore-throat gargle. Later the berries of smooth sumac sporadically were listed as an official medicine in the *The Pharmacopeia of the United States of America*, the first edition of which appeared in 1820. Eventually the Native Americans' uses of berry plants found their way into the Shaker medicinal tome *Druggist's Hand-*

book of *Pure Botanic Preparations* (1873), which recommended, among other things, winterberry bark as a tonic and barberry bark as a purgative.

Once they knew how to use the American berry plants, the colonists simply harvested the woodlands and marshes rather than planting their own. Yet as the seventeenth century progressed and the fledgling settlement societies prospered, the more sophisticated, well-to-do citizenry began to spice up the basic dooryard garden with nonessential yet coveted herbs, berries, and fruit trees. It seems that while the colonists wanted little to do with the motherland, they did miss such delights as authentic British currants and gooseberries. In a fitting symbolic gesture of the marriage of British tradition and American impetuosity, stock sent from England was often grafted onto wild American berry plants. But the berries in the garden were not set out for a great display.

There were, however, those rare individuals who coveted gardens on a par with those of Europe. After inheriting Mount Vernon in 1761, a youthful and ambitious George Washington set about transforming the property into one of the finest pleasure grounds in America. Washington divided the garden into neat sections. In the kitchen garden, called the Lower Garden by Washington, berry and vegetable beds were divided symmetrically and herbs edged these beds. Among his numerous plantings were currants, raspberries, strawberries, and gooseberries. Washington's berry supply was so prodigious that it prompted W. Fitzhugh to write to him in 1798, "You will much oblige me if you can spare me a few garden shrubs, particularly the rasberry [sic] and the gooseberry."

American berry gardens have traditionally boasted a selection of native berries, such as blueberries and strawberries.

Strawberries were the first berries to be cultivated widely on a commercial scale in America, left. Neatly defined rows facilitate maintenance and harvesting.

From the beginning of the eighteenth century to 1850, hybridizing work improved the size, quality, and number of cultivars for American berries. The American native gooseberry was first domesticated in home berry gardens, as was the American black currant. Also at this time, commercial cranberry cultivation got its start in New England.

The invention of canning in the early nineteenth century by the Frenchman Nicholas-François Appert led to the gradual elimination of the casual market garden. Farmers were able to grow much bigger berry crops because the soft fruits could be preserved in large quantities. With the advent of the railroad also in the first part of the nineteenth century, berry growers derived still more benefits. Previously farmers had been forced to sell their goods in the immediate vicinity of

cities. Now they were able to ship berries via express trains from the country to the city. Strawberry growing, in particular, became very widespread during the 1840s and 1850s. During the same period growers in Connecticut made attempts to start a silk industry with white mulberry trees, but the amount of labor required for such a venture precluded its success. Concurrently berry growing became established on the south shore of Long Island, New York, as well as in New Jersey, Delaware, and Chesapeake Bay, Maryland, where blueberries and cranberries produce bumper crops today.

After the Gold Rush of 1849, pioneers began pouring into California. As cities grew and transportation improved, horticultural centers became established on the north and central coasts, where bramble berries and strawberries were culti-

vated. As the population spread into the Pacific Northwest, the Willamette Valley of Oregon became a center for growing such berries as strawberries, trailing blackberries, upright blackberries, raspberries, gooseberries, and currants.

By the turn of the century, middle-class suburban living was coming into its own. Families no longer needed to farm to survive, yet there was a nostalgia for the rural lifestyle as cities began to spread out and commuter railroad suburbs emerged. For its 1888 *Farm Annual*, the Burpee Company of Warminster, Pennsylvania, took heed of its new market, featuring seeds for the suburban produce and flower garden. It also offered a booklet titled *How and What to Grow in a Kitchen Garden of One Acre*. Domestic berry gardening hit a new stride—a trend that continued into the early twentieth century—as suburbanites

Watsonville, California, is one of the leading berry-producing centers in the United States. This old-fashioned farmhouse, with water tower intact, was among the first settlement homes built in the area. Around such humble country homes large expanses of berry fields — many containing raspberries, as pictured — were planted.

Brilliant red raspberries, above, are among the most popular and widely cultivated fruits in America.

virtuously laid strawberry runners and trellised raspberries. They created these miniature farms with the idea that they would enrich their tables and instill a sense of responsibility in their children, who needed a taste of farm living. Thus the modest berry plot that most people are familiar with today took its shape.

Two calamities, the Depression and World War II, also spurred interest in backyard food gardening to defray food costs. The prologue to the *House & Garden Defense Garden Guide* issue, published in January 1942, read: "This year the grim necessities of war focus the gardener's attention on the production of food. Vegetable gardening will again become a popular hobby. *House & Garden* hopes that in addition to this renewed interest in vegetables will come a revived interest in growing fruits, bush fruits especially." The issue went on to list the virtues of berries and explained how their ability to be preserved would lend itself to the important easing of wartime food shortages.

After seven thousand years of known use of the berry by humans—and probably several thousand more undocumented years—it is still a cherished food and one with so much variety that it would be hard to tire of it. The berry connoisseur knows that fresh, succulent berries, harvested at precisely the right moment, are the closest thing to ambrosia that nature has to offer. In this sense, we are not much different from those early northern peoples who enjoyed berries directly from the plant. Such is the magic of berries. They are living links with distant millennia, yet when you taste them, you are truly living in the moment.

Chapter Two

BERRY CULTIVATION

DESIGNING YOUR GARDEN

SELECTING A SITE

Berry gardening can be done in almost any climate. There are varieties of berries adapted to a wide range of climate zones. Berries grow all the way from the Arctic Circle to South America and New Zealand, so varieties are available that can withstand extremes of heat or cold.

In a temperate zone, you will only succeed with exotics like Cape gooseberries if you have a particularly favorable site, or grow them under glass. In a southern region, grow strawberries over currants and gooseberries to avoid sun damage.

Local conditions vary considerably within a few miles, or a few hundred feet of altitude, so it is impossible to make blanket statements about which varieties grow best where. Check with local nurserymen and horticultural specialists for a complete understanding of how soil, local temperature, rainfall and other aspects of climate will affect your berry gardening.

Experts with knowledge of local conditions will also be able to advise on specific varieties of a particular berry that are likely to do well. For example, the 'Blue Crop' highbush blueberry is extremely cold-resistant, whereas the 'Premier' rabbiteye blueberry weathers hot summers well. Similarly, some raspberry varieties are better able than others to tolerate the higher

rainfall in western areas of the British Isles.

Generally speaking, berries require sunny, wind-protected spots. They thrive in an eastern exposure that gets some afternoon shade. When choosing a site for your berry garden, look for a spot that is on a small incline. Since cool air is heavier than warm air, it travels closer to the ground, and a spot on an incline will be less likely to freeze than lower ground. If it isn't possible to avoid frost pockets, shield your berry plants with a windbreak hedge.

You can often improve your site by providing shelter, irrigation, temporary shading and by mulching. Almost any property is a potential site for berries.

At a private home in England, a small berry plot is utilized to its maximum potential, left. Strawberries, bush fruits, and brambles all grow side by side and produce successive harvests.

At Upton House in Banbury, England, red currants flourish on a warm, sunny slope in the kitchen garden, right. Black and white currants also grow in this diversified kitchen garden, along with gooseberries, raspberries, strawberries, mulberries, and loganberries.

LAYOUT

The berry garden is a practical space. Though they're beautiful, berries are meant to be harvested. Organization of rows and easy access to plants are key elements to consider when you plan your berry garden. You have centuries of traditional designs to choose from, whether you want the garden to be a formal, geometric space or a more casual affair with just a row or two of plants. Space bramble and shrub berry plants at least three feet apart within rows and keep the distance between rows at about six feet: wide enough for the plants to spread and for you to walk between them comfortably, even with a wheelbarrow. These distances are for ideal circumstances; you may not have enough space to accommodate the needs of spreading berry plants, so consult your nurseryman for good varieties to squeeze together. If you are training bush berries as a hedge or on a trellis, you should plant them from ten inches to three feet apart.

You will also need to decide on pathway materials—either to border the entire garden or to run between rows. Many gardeners opt for grass, but unfortunately it needs to be cut; even gravel presents problems as weeds can spring up. Bricks, blocks, or stepping-stones are practical options. You can vary textures and colors of these materials if you wish, so that even in winter when the garden lies dormant its borders and walks will be pleasant to look at. Keep in mind, also, that a berry garden can end up looking bare after harvest time. Try to plan attractive beds, paths, trellising, and additional plantings that take the focus off bare, harvested areas.

You do not have to contain all berries in the same part of the garden; some may do better in different areas. You might try strawberries among the annual vegetables (simply because the strawberries must be pulled up and replaced every two or three years) and bush fruits, such as currants and gooseberries, in their own spaces. Try to imagine how different elements of a garden can offset one another. For example, the elderberry's showy flowers often emerge when other berry shrubs, such as currants, are bearing their glistening red fruit. The combination of flower and fruit adds variety.

Whether you have a small or large plot of land, you may wish to draw on the ideas of the grand kitchen gardens of Europe. These gardens, regardless of size, are quite stately, although their plantings contain what have traditionally been considered modest plants—fruits, vegetables, and herbs. In a typical British kitchen garden, there is a sense of fecundity, of bounty. Currants hang like tassels on stocky bushes; pole beans twine up tepeelike sticks trailing profusions of scarlet flowers; dramatic rosettes of globe artichokes stand six feet tall; airy asparagus plants contrast with spires of hollyhock; and brilliant red raspberries complement deep green displays of showy herbs. These varied plantings work together quite harmoniously without seeming haphazard.

If you scoff at the idea of neatly tended rows and crave a more romantic, naturalistic garden in the style of William Robinson, take heart. There is no reason why bush berries—such as blueberries, currants, and gooseberries—can't thrive in the border with other strictly ornamental shrubs. Since they have beautiful decorative qualities aside from their obvious edible virtues, berry shrubs can act as hedges that flower *and* fruit. Used in hedges, bush berries can define garden areas or border pathways. If you prefer a more free-form landscape, consider incorporating random plantings of highbush blueberries mixed with viburnums and other ornamental shrubs, especially azaleas, since they enjoy the same acid soil conditions. This will add interest to your landscape and a deep sense of texture

because you are intermingling different leaf patterns and plant shapes. Similarly, brambles—raspberries and blackberries—can form lush, dense thickets that can define property borders and add sylvan mystery to your garden. Just prune them regularly to keep them in line.

It's also possible to carpet the ground with berry plants. This tradition has been long established: eighteenth-century engravings depict languorous maidens harvesting strawberries from plantings on sunny slopes. Choose alpine strawberries and you'll have a delightful ground cover reminiscent of a forest glen.

In the cleverly designed kitchen garden at this private residence in England, cold frames, greenhouses, and vegetables as well as alternating rows of early-, mid-, and late-season raspberry varieties flourish from summer through autumn.

28

SELECTING PLANTS

One of the dicta that George Washington repeated to the gardeners of Mount Vernon was that there should always be "an Abundance of Everything" in the food garden. This is advice worth considering when selecting plants for the berry garden.

The berry garden need not have a single burst of glory in July. You can choose plants so that you have a steady crop of berries throughout the seasons.

Generally speaking, you can have the following commonly grown berries on this schedule: June-bearing strawberries fruit first; followed by currants, gooseberries, blueberries and bilberries, mulberries, and raspberries throughout midsummer; elderberries, blackberries, Japanese wineberries, and loganberries in late July and August; and late raspberries, Cape gooseberries, and strawberries in September and sometimes as late as October, depending on the season's first frost. Alpine strawberries yield for much of the summer and into fall. Also, if you are willing to put in a great deal of effort, you can have fall cranberries. Huckleberries can fruit anytime from July to October, depending on growing conditions. Of course, fruiting times vary according to the weather in a particular season as well as being determined by broader climatic conditions.

You may not aspire to growing *all* the berries in this list in your garden, but by training some upward and making the best use of space, you could be surprised how many different kinds you can fit in.

Asked which berries are the best to start out with in the backyard garden, berry growers tend to point to the blackberry. One California commercial berry grower says, "Of all the berries that we've ever grown, I think the blackberry is the easiest to get into. Blackberries grow in almost any region, and they do stay in the ground for

ten to twelve years. There always seems to be a home for blackberries." Another grower concurs: "The easiest berries to grow are blackberries. Because they're a northern crop, they seem to thrive in more adverse conditions."

One grower gives sound advice for the home gardener: "Don't plant too many! A few plants go a long way for a household. If you put out two dozen strawberries, that's a nice amount for a family. A half row of blackberries—about eight plants set three feet apart—and a dozen raspberries, set one foot apart, is plenty for a backyard."

Blueberry cultivation requires a certain amount of dedication and perseverance. You must provide your plants with an acidic soil. It's also wise to place netting or a cage around your blueberries to protect them from animal raids.

PLANTING YOUR GARDEN

SOIL TESTING AND AMENDMENT

To get your berries blushing red or glossy black, you must create proper soil conditions. Different varieties of the same berry may prefer different soils, so we will try to avoid making sweeping statements about this topic. In general, fertile, well-drained soil is the key to berry happiness. The soil should be a sandy loam and not be deficient in any important element—such as nitrogen, phosphorus, potassium, and magnesium or the trace elements manganese, iron, zinc, and the like. This means it should have an ample amount of organic material—such as leaf mold, compost, or peat—mixed in and a pH of 5.5 to 7.0. Blueberries and huckleberries like soil that is slightly more acidic, so accommodate them by mixing in additional peat.

Most berries aren't really too fussy about the type of soil they're planted in, but it's a good idea to make the soil as close to ideal as possible. Take a sample to a local nursery to have the mineral content and pH analyzed and fertilize the soil with what it seems to need most for the berries you've chosen to grow.

Give your bramble berries a fertile, well-drained soil and they will bear healthy fruit for a decade or longer.

PLANTING BERRIES

When selecting berry plants, always be sure to obtain virus-free stock. Most berries are self-fertile, which means that a plant is fertilized with pollen from its own flower, so you do not need to plant more than one variety for fertilization to take place. Still, planting several varieties of a particular berry is wise, as cross-pollination usually gives a better set of fruits. (Do not mix red and black raspberries, however, as they may exchange deadly viruses.) For example, if you plant eight blueberry bushes, make sure there are at least two varieties among them.

Berries can be planted in either autumn or spring. Most experts prefer early spring; among other reasons, it allows the plants to establish themselves by the first winter. In hot, dry regions, however, autumn planting prevents damaging vulnerable plants during the summer months. Currants and gooseberries are exceptions to the spring-planting rule. These plants start growth very early in the spring, so they must be planted in late autumn or winter.

Bramble berries, such as blackberries and raspberries, are planted in the spring. They are generally sold by nurseries "bare root," which means that they have been removed from the ground in winter in a dormant stage. Nurseries will usually tag a line on a bare-root plant to show at what level of soil it was planted before being uprooted; it should be planted at the same depth. Cut back the plants so that two or three buds rise above the soil and then cover these with several inches of mulch in a mound formation. If you intend to stake the plants for support, add the stake at this time.

Bush berries—blueberries, currants, and

Bramble berries, above, are generally planted in spring. These 'Himalaya' blackberries grow at the Royal Horticultural Society's Garden at Wisley and are trained on wires and stakes.

If you obtain certified virus-free stock from a reputable nursery, you should have no trouble cultivating glossy, healthy berries such as these, right.

gooseberries—are generally purchased in containers, although sometimes they are also sold bare root. Plant them about an inch lower than the depth they reached in the container—or as indicated on the bare-root plant. Be sure to add moistened peat or compost to the planting hole to stimulate root growth. With currants and gooseberries, plant young bushes with no more than six branches at the maximum and keep well mulched.

Elderberries, which are bush berries, are grown according to a variety of methods, most commonly by seed or by transplanting them bare root. Since it is unclear whether the elderberry needs another plant for pollination purposes, play it safe and plant at least two varieties near one another to ensure healthy berries. These intriguing berries are at home in well-drained soil and a sunny location.

Some of the other berries discussed in this book—cranberries, lingonberries, Cape gooseberries, mulberries, and rowan-berries, as well as all of the ornamental berries—require special growing conditions. In some cases, as for cranberries and lingonberries, cultivation is extremely complicated, and the berry gardener eager to grow these types should consult specialized sources. The other berries, while not difficult to grow, each demand different climatic and soil conditions, too lengthy to include here but descriptions of which are available in any gardening encyclopedia or from your local nursery.

PLANTING STRAWBERRIES

Strawberries aren't particular about soil, but they're extremely idiosyncratic about how they're planted: they must be set at the correct depth. As you place the plants in the ground, be sure to fan out the roots. Set the strawberry plants as far as they will go into the soil without covering the tip of the crown—which means new leaf buds should sit directly on top of the soil. Be careful that the crown is positioned precisely on top of the soil; if it's set too deeply or too shallowly, the plant will founder. Finally, pack soil around the plant firmly and water the strawberry plant generously.

To ensure vigorous fruit from newly planted strawberries, remove blossom stems from the plants as soon as they appear during the first flowering season.

There are three methods for planting strawberries, each with its own merits. In the hill system, favored by gardeners who must make do with a small patch, plants are set in double or triple rows, with about a foot between plants and rows. Leave at least a fourteen-inch walkway between groups of rows. When plants develop runners, pinch them off. Large berries result from this time-consuming method, which requires constant tending to keep runners under control. Some gardeners allow a few runners to develop so that they can create replacement plants.

At the other extreme is the matted-row system, a favorite among weekend strawberry gardeners. Plants are set one and a half to two feet apart in rows separated by at least three feet. Virtually all runners are allowed to develop around the "mother plant." Berries grown using this method are smaller than those of the hill system, but, of course, this system does not require as much effort.

The compromise is the spaced matted-row system, which uses the same layout as the matted-row method. When the mother plant develops runners, limit them to four and manipulate the runners into tidy cross formations. It's a little more difficult to maintain than the matted-row technique but easier than the hill system. The berries will be larger than those produced by the matted rows but smaller than the fruits of the hill system.

The life of the cultivated strawberry plant is sweet but fleeting. It tends to decline quickly in vigor and yield over the years. For optimum berry production, plan on tilling under your old patch and starting a new one after the second harvest. If you want to attempt to cultivate a perpetual strawberry bed, be scrupulous about weeding and search out and destroy the rangiest-looking plants after each harvest. It is imperative to continue to introduce new, virus-free plants into this "permanent" bed from time to time.

Alpine strawberries do not produce runners. This tidy habit makes them ideal for borders and as ground covers. It also means they must be planted in an entirely different fashion from cultivated strawberries. Germinate the seeds indoors in late winter and plant the seedlings in early spring, about a foot apart. These tiny berries won't make you wait long to enjoy them: plants started in February should bear fruit by June. A patch will last for about five years.

Maintenance of the traditional large-fruited strawberry plot involves removing old plants and planting new ones every two years or so, above. By contrast, alpine strawberries, right, can be left in place for about five years.

A well-tended strawberry patch located just outside the back door, left, provides easy access to berries and makes it possible to enjoy fresh strawberries in season when the impulse strikes.

GROWTH METHODS

CONTAINERS

Container plantings of berries are elegant and practical. With containers, you have the option of instantly redesigning your berry garden, moving berries to more congenial spots, and creating colorful accents on your property. Urns of strawberries framing a walkway are regal and utilitarian. Although pots and tubs of strawberries are familiar sights, containers brimming with other berry fruits are not only feasible but can be strikingly original garden accents.

For urban and many suburban gardeners, it is often necessary to squeeze productivity out of small spaces. This can be accomplished through container plantings. A container of blueberries can bring gorgeous autumn color to a city terrace. Currants will flourish in a terra-cotta pot, flowering beautifully in spring and then fruiting abundantly in summer. Brambles such as raspberries or blackberries lend craggy rusticity to patios when planted in large wooden tubs.

There are several methods of constructing berry cages. This grower opted for a rustic style, opposite page. Casual plantings of flowers harmonize with this informal approach to gardening.

Gardener Neil Hewertson constructed a larch berry cage of octagonal design to house berries at Stowell Park, a private estate in England, left. Paths within the cage fan out in a wheel shape interplanted with gooseberries and black, red, and white currants. This structure is attractive enough to hold its own in a garden that already houses an ancient orangery.

PRUNING AND TRAINING

The purpose of pruning edible berries is to stimulate end buds to full development. Some, more laissez-faire berry gardeners get by without pruning, putting their faith in the greater wisdom of nature and pointing to the abundance of berries yielded from un-pruned and untrained wild bushes and brambles. While this approach to pruning is appealing, it can defeat the purpose of having a berry garden to begin with. Most of us don't have a forest at our disposal in which to gather berries, and so to harvest high yields of healthy berries from fewer numbers of plants we need to control the fruiting process.

Pruning and training are means to the same end in edible berry gardening and are practiced in tandem. Most bramble, or cane, berries require some kind of support for optimal fruiting. Usually they are trained onto a trellis of two to four wires spaced evenly up to six feet high; the wires, in turn, are supported by posts every ten feet or so apart.

Bramble berries bear flowers and fruit on erect or trailing canes grown the previous year. In the case of blackberries, black raspberries, and purple raspberries, the canes fruit once and then must be removed immediately after the crop has been harvested or during the winter after harvest. As the plants fruit, they also sprout new canes, which should be pruned back to between four and six in number. After the bearing canes are removed, the new canes should be tied to a trellis wire. Especially with blackberries, these new, fast-growing canes will have to be cut back regularly during the summer to about a foot each; toward the end of winter, remove frost-damaged growths. An experienced home berry grower sums it up as a four-step process: "Prune out what's fruited; during the growing season, pinch out the growing tips; during the dormant season, thin out the canes; and shorten the lateral branches to a foot and a half."

Everbearing red or golden raspberries will produce an autumn crop of berries on the tips of the new canes. To get a good crop of autumn berries, remove the old canes that fruited in summer and train the new canes to a post or wire. After harvest, prune these back so that the lower parts can bear fruit the following summer. One berry gardener makes this process simple to remember: "Prune out what's fruited, thin out canes that are one year old, and head back the canes you've saved."

Another way to train brambles is with a V-shaped wooden support. Fruiting canes can be trained onto both sides of the V, and new canes can grow directly up the middle. This method is quite picturesque; it also allows for a large harvest because brambles can be planted closely together and then separated on the V. You might also construct a trellis using wooden poles instead of wires to create a rustic mood in your berry plot. Adventurous garden designers may enjoy training the canes of trailing blackberries in a fan shape or weaving them around horizontal wires in attractive spirals. Brambles can also be trained against a wall or fence, a practice often employed in many of the important kitchen gardens of Europe.

There are differences of opinion among berry growers as to what materials are best for use as stakes. Perry Kozlowski of Kozlowski Berry Farm in Forestville, California, favors iron stakes for training berries. He finds that "grape stakes," the same type used in vineyards, are effective and sturdy and are far more cost-effective and durable than wood. Nita Gizdich of the Gizdich Ranch in Watsonville, California, rejoins, "We tried metal poles, and when we got a windstorm, it put them right down. Now we're back to two-by-two wooden stakes." The Gizdich farmhands apply a wood preservative to the posts, and they are used indefinitely.

Inspiring examples of trained berries can be seen at the Royal Horticultural Society's Garden at Wisley, Surrey. Upon entering the Fruit Collection, the visitor encounters a long, grassy walkway framed by two parallel fences on which are trained trees and berries. The effect is similar to that of a pleached allée—a hedgelike, interwoven wall of trees and shrubs common at grand country houses.

Bush berries do not need to be trained. Some berry gardeners do train them on trellis wires, as with bramble berries, or as "pillars," on a single, central stalk. Bush berries most definitely require pruning. For currants and gooseberries, wood that is four or more years old should be removed in winter; this allows a new scaffolding of branches to form. Cut back the bushes to about six branches with five buds each. Even if the bush is immature and hasn't developed six branches yet, you should do this with its branches, pruning them back to about three buds each. Pruning is especially important if a bush bore small fruit in the summer.

For blueberries, the procedure is even simpler: cut away dead wood in winter; light pruning is usually sufficient. As with other bush berries, do not allow wood over four years old to remain on the plant; of course, you'll have to pay careful attention to the process of the plant's growth progress over a time period of years.

Pruning is not nearly as crucial for ornamental berry plants as it is for the edibles. Prune your ornamentals to give the plants attractive shapes and to encourage flower-

ing. Young flowering trees benefit from light pruning up to the time that they reach the flowering age; thereafter, pruning has little effect on flower production. For shrubs, removing dead branches at the bottom of the plant and thinning out older wood is effective. Prune spring-flowering shrubs, like viburnum, after flowering. Late-flowering shrubs, like roses and snowberry, should be pruned only in winter or early spring.

Instead of thinking of pruning ornamentals as drudgery, try viewing it as a means of artistic expression. Outfitted with pruning shears and an eye for the artistic, the aesthetically minded berry gardener can pleasingly shape the home landscape. As garden writer Vita Sackville-West once wrote, "A bit of judicious cutting, snipping and chopping here and there will often make the whole difference. It may expose an aspect never noticed before . . . a coloured clump in the distance, hitherto hidden behind some overgrown bush. . . . It is . . . like being a painter, giving the final touches to his canvas: putting just a dash of blue or yellow or red where it is wanted to complete the picture and to make it come together in a satisfactory whole."

In the walled kitchen garden at Glebe Court, the private residence of Lady Hobhouse in England, a trained blackberry archway provides an air of sylvan mystery as well as baskets brimming with succulent fruit.

CULTIVATING A BERRY PATCH

MULCH

Mulch can be any of a number of materials—peat moss, bark, straw, pine needles, leaves, or black plastic—employed to protect an area by keeping soil moist, preventing the germination of weeds, and in general reducing maintenance. Simply put, mulch is an effective means of protecting berry plants from the elements, and it takes little time to do.

Organic mulches have the added advantage of enhancing organic content in soil. (Note, however, that mice can be a problem with straw mulch.) Black plastic mulch is often used by commercial berry growers, who favor it because it is inorganic and therefore permanent. It's awfully hard for a weed to get a roothold under black plastic, so it's a real work saver. However, in temperate climates with ample rainfall, black plastic mulch may allow too much moisture to go to the plants because of lack of aeration.

Mulch also provides winter protection for plants. The poet Tusser offered the six-teenth-century English cottage housewife some advice about mulching that is still applicable today:

> If frost do continue, take this for a law
> The strawberries look to be covered with straw.

Some people claim that the strawberry got its name from its association with straw mulch. The plant is closely associated with mulch because of the necessity of protecting its early-blooming blossoms from frost.

Professional and serious growers unanimously support mulching for the home berry gardener, preferably with an organic mulch. One berry gardener explains its importance: "Berries, more so than any other plants, respond to mulch because they're plants of northern climates, are shallow-rooted, and like moisture in the soil." Another grower has a cost-effective method for mulching: "When our raspberries and blackberries are pruned, their trimmings are chopped up and mulched back into the ground."

Do not skimp on mulch if you want your berry plants to be strong and healthy. Organic mulches add nutrients to soil while providing cold-weather protection.

PROTECTION FROM BIRDS

It's a good idea to use netting or cages to shield berries from birds and other hungry beasts. Cages are more expensive but of course are permanent. Netting tends to work better with strawberries, supported in a tentlike fashion with stakes, although I have seen it used on bramble and bush berries.

Another possibility is to create a scarecrow. In Finland, these are often seen in berry patches. They are not, however, the cheerful freckle-faced types Americans are used to seeing in fields. The Finns get straight to the point: scarecrows are often bizarre hybrid creatures wearing old coats on their backs and crow corpses on their faces. This seems to be effective, if not appealing to the eye.

Some berry growers claim that birds are only after the moisture content in berries and won't eat them if you place a trough of water in the garden. This seems highly unlikely but is certainly worth trying.

A berry grower in upstate New York takes special precautions to protect his blueberries from the birds by encasing them in berry cages of wood and chicken wire.

WATER

Commercial berry growers use self-running irrigation systems that ensure measured dosages of water to berries, but a sprinkler system is sufficient for your purposes. It is especially important to water during dry summer months. It's usually pretty easy to tell when plants need watering: they wilt, and some of the berries may have a sunburn, which gives them a withered appearance. Highbush blueberries, in particular, require an ample water supply.

Many experts suggest watering plants once a week, in the morning or in midafternoon, to give the berries a chance to dry before nightfall so that berry rot does not occur. Give berry plants a complete soaking after harvesting.

At this humble cottage garden in England, daily dousings with a watering can ensure healthy, vigorous bramble growth.

Chapter Three

EDIBLE BERRIES WITH RECIPES

INTRODUCTION
THE EDIBLE BERRY GARDEN

For most of us, the very word *berry* conjures up childhood memories of sun-drenched days spent picking and consuming these small fruits with abandon. We remember the aroma of strawberries in the summer heat, the sensation of raspberries bursting in the mouth, and the tangy sweetness of blackberries picked from roadside hedgerows. In his *Remembrance of Things Past*, Marcel Proust's childhood memories are unlocked when he tastes a type of cake served in his youth. Like Proust's *madeleine*, berries tend to have pleasant associations and evoke the feeling of childhood, barefoot summers, and simpler times. Perhaps that is why, aside from their taste, people enjoy berries so.

Berry growers attest that there are few things so gratifying as harvesting the fruits from your own garden. Not only will you find the berries more fresh and delicious than store-bought varieties, but you will also derive immeasurable pleasure as you observe the changes in the garden throughout the seasons. The springtime flowers of berry plants augur months of summer bounty: currants dangle delicate pink petals, blackberries lure bumblebees with clusters of white blooms, strawberry flowers burst forth in a carpet of white and yellow. Come early summer, strawberries gleam like rubies in the sun. Later, glossy black, red, and white currants festoon the berry garden. Gooseberry bushes sprout plump, inviting fruits. As autumn approaches, fall-fruiting raspberries of brightly contrasting tones—magenta, black, golden—renew color in the garden, while blueberry foliage begins to turn to a lovely reddish hue.

Certainly you don't have to grow berries yourself to enjoy them. Aside from the garden varieties, there are also wild edible berries. There are few surprises so pleasant as coming upon by chance a patch of uncultivated berries, sitting there for the taking. Remember, however, that some kinds of berries are not edible so before eating any berry taken from the wild you must consult a region-specific field guide to make certain you know the kind of berry you are picking and that this berry is in fact edible, rather than poisonous. Bramble berries—such as blackberries and raspberries—bilberries, huckleberries, blueberries, rowanberries, and elderberries are among the plants that usually grow uncultivated in North America and Europe, although not all species occur on both continents. In Scandinavia, where wild berries are so prevalent that you cannot avoid stepping on them in the woods in summer, lingonberries, cloudberries, and Arctic brambleberries are the prizes of the berry forager's outing.

Strange as it may seem to us today, there was a time when berries were not regarded primarily as food. As late as the nineteenth century in the British Isles, Scandinavia, and elsewhere in Europe, berries held significance in folklore, and many powers were ascribed to them. They were regarded as omens in dreams, talismans that protect against evil spirits, and cures for whooping cough and warts, among many other ailments of the times.

Acquainting yourself with this elaborate lore will undoubtedly give you a newfound respect for berries. Perhaps it will spur you to create a berry garden or become an expert on a particular genus. But one thing is certain: the marvelous history of berries will inspire you to bring them more frequently to your table.

Blackberries fruit prolifically against a shed in Great Rollright, England. These plants were propagated from tips of root cuttings taken from plants of a grand manorial garden across the road and required little care to become so robust.

ARCTIC BRAMBLEBERRY
Rubus arcticus

For those who wish to plunge into the realm of berry exotica headfirst, there is the Arctic brambleberry. This berry truly has an untamed spirit, so unless you live in Scandinavia, to which it is native, it would be folly to try to cultivate it. Perhaps it's most gratifying simply to appreciate its beauty in photographs and wonder at its rarity.

The reputation of the Arctic brambleberry now extends well beyond Scandinavia. Although Scandinavians have been gathering this rare and delicate berry from the woodlands for centuries, its use outside of these countries has been known for only a hundred years or so. Finns proudly tell the story of a Turk who wanted to impress his guests with a banquet. For days and nights he pondered what delicacy he could serve. Eventually he selected the ultimate beverage—a Finnish liqueur made from the Arctic brambleberry.

The Arctic brambleberry is not so easy to come by even in northern countries. When visiting a farmer friend who lives in Finland near the Soviet border, I asked where I could find this legendary berry, known to the Finns as *mesimarja*, which translates as "honey berry." He scratched his beard and looked off into the wilderness. Finally he came up with a plan. We went to visit a family that had tried to cultivate the berry. After much loud, enthusiastic conversation (the *mesimarja* always seems to inspire

Finns to talk loudly) we put on our boots and sallied forth into the family's patch.

But the *mesimarja* plants, spaced neatly like strawberries in the family's plot, were simply not fruiting, although they were in season. My farmer friend nodded his head knowingly. The sensitive honey berry, which thrives in semirocky woodland soil, defies cultivation and some years produces little or no fruit.

After much consideration my friend thought of a new tack. We got into his truck and drove on side roads near the Soviet border that were dotted with signs warning against entry. After driving for a time with a careful eye on the side of the dirt road, he finally shrieked, "*Mesimarja!*" and stopped the vehicle. He gently brushed aside other roadside plants at the edge of a great Finnish-Soviet forest and revealed a few tiny honey berry plants—one of the most precious sites a berry forager can find.

The Arctic brambleberry resembles a raspberry but actually is related more closely to the blackberry. Its odd, stepped conic shape gives it a crafted look, and it grows, an opalescent red jewel, in solitude among wildflowers, lingonberries, and wild raspberries. But it is the taste that inspires berry pickers to search for it tirelessly. It has the familiar sweetness of a raspberry but is somehow more inspired and otherworldly—a berry that could grace the table in the hall of Valhalla.

The reclusive Arctic brambleberry betrays its hiding place among wild plants of the Finnish woodland through its vivid red coloration. Finns turn the berries into heavenly desserts and liqueurs.

48

Brambleberry Sorbet

Should you be so lucky to track down the Arctic brambleberry, here is a wonderful way to put it to use. Nestled in a hazelnut crown, brambleberry sorbet is an elegant, impressive dessert. Alternatively, blackberries could be used.

(Serves 4)

Sorbet:
*225g/8 oz fresh or frozen
    brambleberries*
150ml/¼ pt water
6 tablespoons caster sugar
5 tablespoons whipping cream

Sauce:
*50g/2 oz fresh or frozen
    brambleberries*
3 tablespoons sugar

Hazelnut Crown:
50g/2 oz sugar
40g/1½ oz ground hazelnuts
2 tablespoons plain flour
2 egg whites

Sorbet: Make a purée of the brambleberries; you should have about 150ml/¼ pt of purée. Mix the water and sugar. Bring to the boil and cool to a syrup. Mix in the brambleberry purée.

Using an ice cream maker, follow the manufacturer's instructions for processing sorbets, until almost frozen. Add the cream and continue to process until you have a fluffy mousse. If you do not have an ice cream maker, pour the mixture into a bowl and place it in the freezer. When it starts to solidify, whisk it with an electric beater until smooth. Repeat a couple of times while the mixture is freezing (it takes 3 to 4 hours). Remove the brambleberry mixture from the freezer before it is solid and add the cream. Return it to the freezer. Beat the mousse once again just before serving.

Sauce: Mix the berries and sugar in a saucepan and bring to the boil. Rub the mixture through a sieve. Cool.

Hazelnut Crown: Cut out a six-pointed star (12.5cm/5 inches in diameter) from a thin piece of cardboard. Discard the star, but keep the cardboard from which it was cut so that you have an outline of the star. Butter a baking sheet.

Preheat the oven to 200°C/400°F/gas mark 6. Place the cardboard on the baking sheet. Mix the crown ingredients and pour the mixture into the star shape. Repeat to make four stars. Bake for 8 minutes, until golden brown. Remove the stars one by one before they cool and place them in small glass bowls to form the cuplike shapes.

To serve, divide the sorbet among the crowns and top with sauce.

BLACKBERRY

HIMALAYA BLACKBERRY
Rubus procerus

EVERGREEN BLACKBERRY
Rubus laciniatus

HIGHBUSH BLACKBERRY
Rubus alleghteniensis

Should you happen to wander not far into any British wood, you will likely encounter a blackberry thicket. Rising perilously above your head, the arching canes intertwine in a thorny maze resembling a collection of enormous croquet wickets. The sight of this impenetrable bramble—groaning with lustrous berries under the dark forest canopy—is striking yet also a bit daunting, calling to mind folktales of moss-clad woodland spirits and witches hidden in briars casting spells on passersby.

To the medieval peasants of the British countryside, the blackberry—or "bramble," as it is known in England and Scotland—possessed powers as formidable as its ferocious thorns. Hence a rich lore of superstition surrounds the blackberry. The blackberry could cure, it was believed, but it could also curse. According to a Cornish belief, the first blackberry spotted growing in the wild each year will banish ugly warts.

In Devonshire, it was (and perhaps still is) thought that creeping under a bramble three times from east to west would cure boils and "pinsoles." The curative bramble has to form a natural arch, as blackberry canes often do, and it is considered a particularly lucky plant if the root tips of the arch touch the property lines of two different landowners. Crawling through a bramble is thought to relieve the dreaded

whooping cough in children, but the adult supervising the procedure must chant, "In bramble, out cough; Here I leave the whooping cough."

The flowers and leaves of the blackberry are also considered powerful. If the plant blossoms in early June, it is taken as a portent of an early harvest. In Cornwall, countryfolk would apply blackberry leaves to scalds while repeating a charm.

But blackberries can also heap trouble on the unsuspecting. In Britain, there is a folktale of a man who "jumped into a bramble bush and scratched out both his eyes," quite enough to instill an enduring

Wild blackberries, above, are free for the taking and have a magical sweetness. Cultivated blackberries also have much to recommend them. Trained on wires and posts, evergreen blackberries, left, bask in the early autumn California sun.

fear of this plant in a child's mind. It has long been considered unlucky to pick blackberries past October 10, Old Michaelmas Day. It is said that Satan puts his cloven foot on the blackberries on that day, cursing anyone who is tempted to partake of them.

The blackberry also enters the realm of dreams. If you dream of journeying through bramble-covered places and being pricked by the thorns, then "secret enemies" will conspire against you; if you bleed, death could be imminent. However, should you pass through the dream brambles unscathed, you will experience triumph over your foes.

The voluminous lore surrounding this enigmatic berry is not surprising when you consider its cleft nature. The fruits are raven black, the flowers pink or immaculate white; the thorns pierce, but the berries entice. The blackberry also grows profusely in British roadside hedgerows, where it takes on a less formidable appearance. The garden writer William Robinson, that champion of the natural hedgerow, opined that it is "very interesting . . . to observe the differences between some of the sub-species and varieties of blackberries, and the beauty, both in fruit and flower, of the family." Robinson's words no doubt prompted many a gardener to let the bramble into the home landscape—one potential explanation for the existence of so many blackberry thickets in England.

The species seen frequently in gardens in the United Kingdom is Himalaya blackberry, *Rubus procerus*, a strong, prolific grower. It is another species, however, that inspired Dylan Thomas to evoke gloomily the dark side of the blackberry in "Poem on His Birthday":

Dark is a way and light is a place,
Heaven that never was
Nor will be ever is always true,
And in that brambled void,
Plenty as blackberries in the woods
The dead grow for His joy.

In the United States, the evergreen blackberry, *R. laciniatus*, and the highbush blackberry, *R. allegheniensis*, are prevalent both in the wild and in cultivation. The evergreen blackberry has gorgeous deep green leaves and semitrailing canes that can be trained in languid loops in a trellis. The highbush sprouts stout purplish-red canes that can grow as long as six feet. It thrives in forests, but it also favors clearings and roadside thickets. The highbush blackberry's lovely white flowers blossom from May through June; the thimble-shaped berries follow, ripening to a deep, glossy black by late summer. This may have been the species whose roots helped cure the Oneida Indians of dysentery one year while the colonists suffered. It is a widely adapted species, growing happily from New Brunswick, Canada, to northern Georgia. More southern-oriented species of blackberries are popularly known as dewberries. *R. trivialis*, southern dewberry, is the best-known species. It festoons forests and plains with delicate pink or white flowers as soon as winter's chill subsides and rewards foragers with elongated clusters of berries from April through June.

Watch for blackberry flowers in the spring, not only to appreciate their beauty but also to take note of future blackberrying spots for summer. Likely locations are roadsides, railway embankments and the edges of woodlands. For information on other relatives of the blackberry, see the section on *Rubus* varieties (page 138).

Whether or not you believe the lore of the blackberry, you cannot help falling prey to their edible charms.

British cottagers favor the blackberry for their home plots, left. Few berries offer such a profusion of fruit for so little trouble.

Sautéed Veal Chop with Blackberry Sauce

This elegant main course combines the tangy blackberry with veal. Although it looks sumptuous, the dish takes only 15 minutes to prepare.

(Serves 1)

1 veal chop
salt and pepper to taste
50ml/2 fl oz red wine
50ml/2 fl oz chicken stock
25g/1 oz blackberries (preferably fresh, although whole berries in syrup will work)
10g/¼ oz unsalted butter

Season the veal chop with salt and pepper. In a hot frying pan, brown one side of the veal, lower the heat and cook the other side until done. Remove the veal and put on a serving plate.

Return the frying pan to high heat. Pour the wine into the pan and reduce by half. Add the chicken stock, bring to the boil, and allow the mixture to reduce for a minute or two.

Add the blackberries and butter. Swirl the pan around so the butter melts evenly. Season with salt and pepper to taste. Pour the sauce directly onto the veal.

Tangy Blackberry Chutney

Chutney is an elegant way to enjoy blackberries as an accompaniment to a main-course dish. This beautiful mixture can be served with pork or game. A sparkling glass of beer and a crusty piece of fresh bread complete this hearty meal.

(Serves 8)

1 medium onion, chopped
1 tablespoon finely chopped garlic
3 tablespoons olive oil
1½ teaspoons salt
50ml/2 fl oz white wine vinegar
100–150g/4–5 oz light brown sugar
1 teaspoon ground cumin
¼ teaspoon cayenne pepper
½ teaspoon ground cinnamon
½ teaspoon nutmeg, preferably freshly grated
550g/1¼ lb sliced fresh peaches
100g/4 oz fresh blackberries
65g/2½ oz walnut halves

In a large frying pan, sauté the onion and garlic in olive oil over low heat. Add the salt, vinegar, sugar, cumin, cayenne pepper, cinnamon and nutmeg, blending well.

Add the peaches, blackberries and nuts. Continue cooking over low heat for 12 to 15 minutes, until the fruit is lightly cooked. Allow to cool. For the best taste, cover the mixture and refrigerate the chutney overnight before using it. It keeps for months when stored in a covered jar, as do commercially available chutneys.

Brambles and vegetables grow side by side in this cottage garden adjacent to farm-lands in the British countryside.

BLUEBERRY

HIGHBUSH BLUEBERRY
Vaccinium corymbosum

BILBERRY
Vaccinium myrtillus

Smooth, perfectly round, and dark blue with a slight whitish bloom, the modest, unassuming blueberry would seem to be the most easily identifiable of the berries. Yet this winsome chameleon, a member of the genus *Vaccinium*, has an entourage of extended family and near-relations so similar in appearance that identifying any species with certainty can be next to impossible. To muddle the situation further, the blueberry brood grows in several guises—towering overhead or forming a low dense carpet, termed a "barrens."

Native North American blueberries are referred to as either highbush or lowbush. The highbush types, principally *V. corymbosum*, have large, sweet berries; these are the blueberries that grow in cultivation. They can reach heights of twenty feet, creating their own closed societies in deciduous forests, hillsides, or acid bogs.

The lowbush sorts, such as *V. angustifolium*, dote on sandy or rocky acid soils, are quite hardy, and have a wide range. These petite plants only grow from eight inches to one and a half feet high. The small, tart fruits, which vary from cadet blue to a purplish black, inspire pickers to trek deeply into the woods to procure them. Lowbush plants have a spirit that can't be broken. These finicky berries strongly resist cultivation in the home garden but thrive in the wild. Look for their greenish-pink urn-shaped flowers dotting the ground in early spring, before the plant's leaves appear.

Confusion ensues when we introduce the bilberry, or whortleberry, which is the blueberry's closest European relative. It is a favorite of berry pickers in Northern European climes and is quite similar to the blueberry in appearance and taste. The key to differentiating between the blueberry and the bilberry is by the flowers, which are urn-shaped and star shaped, respectively. The fruits of the bilberry are, as a rule, darker than blueberries, and grow singly on axils, as opposed to blueberry clusters.

Mature blueberry shrubs, left, reward the gardener with fruit-laden branches in summer. One plant can yield as much as twenty-five pounds of berries. At this upstate New York farm, blueberries are in full tilt by mid-July, but you can have berries all summer long if you plant early-, mid-, and late-season varieties.

Bilberries grow wild in dense carpets on the floors of Finnish forests, above. It is a favorite of berry pickers in Northern European climes and is quite similar to the blueberry in appearance and taste. The first American colonists confused the native lowbush blueberries with the bilberries of the Continent.

But the bilberry suffers from multiple personalities even in its native lands. In rural Britain, this country cousin might go by the names *huckleberry* and *huddleberry* (in Sussex); *hurts* (Devon), and *whimberry* (central England and Wales). In William Robinson's *The Wild Garden*, the author casts a broad net when discussing the native wild berries of England and lists bilberry and whortleberry as two entirely *different* berries.

The Finns give the bilberry an evocative name, *mustikka*, which is often translated as blueberry or huckleberry, just to keep things sufficiently confusing. It grows with abandon in the woodlands, intermixed with lingonberries, forming colorful patterns on the forest floor. In Australia, the fruits of the myoporum and blueberry ash trees are known as blueberries, further adding to the confusion and consternation of blueberryphiles.

Not surprisingly, the blueberry also confounded the first American colonists. John Josselyn, who diligently cataloged the flora of the New World, reported upon North American native blueberries in a most terse manner. Of "Billberries," he said, there are "two kinds, Black and Sky Coloured, which is more frequent." Josselyn was probably referring to blueberries and huckleberries, another blueberry look-alike that has fueled the nomenclature jumble but is not even a member of the *Vaccinium* genus. Although Josselyn misidentified blueberries, he found that they "are very good to allay the burning heat of Feavers and hot Agues, either in Syrup or Conserve." Native Americans were able to derive profit from the colonists' immediate weakness for blueberries; they sold them sun-dried specimens by the bushel so that the settlers could add color and piquant flavor to their meals even during the chilly New England winter.

One North American species common to the Pacific Northwest and Canada, *V. membranaceum*, is sometimes called a bilberry, but this is probably because British Columbians named it feeling nostalgic for

the indigenous berries of their homeland.

Other impersonators have sought to sully the blueberry's sweet, wholesome image. The twinberry (*Mitchella repens*), which grows with abandon in the Pacific Northwest, is one such doppelgänger. It closely resembles the blueberry, although it has a reddish tinge, and is edible but is actually not very pleasant-tasting. Despite the appropriateness of the twinberry's name as far as blueberries are concerned, the berries are so named because of their habit of producing fruit *à deux*, in a twin-like union. A common childhood prank is to mix the twinberries with blueberries and offer them to the unsuspecting. The victim of the sour twinberry forever curses the blueberry and all its kith and kin.

Such discredits to the blueberry name are more than made up for by those who have paid homage to this cerulean fruit. Thoreau wrote enthusiastically (and almost smugly) about how he was able to feast for several days on blueberries gathered during one day's picking. Robert Frost named a poem after blueberries and even touched on their habit of springing up in fire-ravaged areas, writing that once all vegetation is cleared, "they're up all around you as thick/And hard to explain as a conjurer's trick."

Immature blueberries dote on sun, opposite page. These berries are obviously not ready to be eaten, but even when blueberries look table-ready they may still be in the ripening process. Always taste blueberries before harvesting them to ensure they've achieved their peak flavor.

Even though it flourishes in the wild, the blueberry can also be tamed and made to produce flavorful berries in the home garden, as shown at left.

Blueberries have an extensive depth and range of glorious color that make them almost too picturesque to pick. If you like the idea of a cool blue summer garden, why not fashion a landscape in shades of blue to complement a blueberry plot? To establish a foreground, plant the blue-flowered woodruff (*Asperula orientalis*), blue lobelia (*Lobelia siphilitica*), and California bluebell (*Phacelia campanularia*)—all summer- or autumn-blooming plants that will enliven the garden over several months' time. For a little more height, add wild monkshood (*Aconitum uncinatum*), blue phlox (*Phlox divaricata*), and Quaker bonnets (*Lupinus perennis*), bearing in mind that the blueberry plot must have a more acidic soil than that of its neighboring plants.

Ericaceous plants (heath family) in general mix well with the blueberry because they thrive in the same soil. Especially nice in the blue garden is the gray-leaved heather, *Calluna vulgaris hirsuta*, whose restrained coloration harmonizes with the more pronounced tones of the blueberry.

Blueberries are hardy plants, and they demand the same strong, unyielding constitution of their growers. The blueberry gardener should have a streak of Yankee obstinacy and a persevering nature: the plants don't produce a large crop until reaching their sixth year. If you simply must have home-grown berries without too much delay, plant a two- or three-year-old plant among the others to lessen the wait. Once established, bushes will be faithful fruiters, lasting for twenty-five years or longer. There are early, midseason, and late-bearing varieties, so the blueberry zealot can be sated throughout the warm-weather months.

Blueberries are cherished by some gardeners as much for their luxuriant foliage, which turns reddish in autumn, as they are for their sweet summer fruits.

Cold Blueberry Soup

Although cold fruit soups are associated with Scandinavia, this recipe comes from California. In summertime, cold soup is the perfect light meal, and this recipe is just right for an alfresco meal on the patio. Its vivid blue colour and delicate scrolls of soured cream make it as beautiful as it is tasty. To make the lacy shapes with soured cream, pipe it through an icing tube. If you want the meal to be more casual, simply add a dollop of soured cream to the centre of the soup.

(Serves 4)

900g/2 lb blueberries or bilberries
750ml/1¼ pt water
100g/4 oz sugar
½ teaspoon ground cinnamon
¼ teaspoon ground allspice
½ teaspoon salt
juice of 1 medium lemon
50–120ml/2–4 fl oz crème de cassis
1 tablespoon blueberry or bilberry
  vinegar
soured cream
fresh tarragon (optional)
strips of candied orange peel (optional)

Place the blueberries, water, sugar, cinnamon, allspice and salt in a large saucepan and cook over high heat for 3 to 5 minutes. Allow to cool. Add the lemon juice, crème de cassis and blueberry vinegar. Process in a blender or food processor fitted with a steel blade until smooth. (The mixture may need to be strained to achieve a really smooth texture.) Refrigerate for at least 30 minutes, until chilled. Garnish with soured cream, fresh tarragon and orange peel.

Best Blueberry Muffins

The proliferation of delectable wild blueberries (known as bilberries here) throughout New England has inspired American cooks to use them in just about every food imaginable—from pancakes to puddings to sauces for fish. This blueberry dish is a quintessential New England food: blueberry muffins.

(Makes 8 muffins)

> *150g/5 oz unbleached flour*
> *100g/4 oz sugar*
> *2 teaspoons baking powder*
> *¼ teaspoon salt*
> *grated zest of 1 lemon*
> *150g/5 oz blueberries or bilberries*
> *1 egg*
> *2 teaspoons soured cream*
> *50ml/2 fl oz milk*

Preheat the oven to 180°C/350°F/gas mark 4. Mix the dry ingredients with the blueberries in a mixing bowl. Add the egg, soured cream and milk. Stir until just moist. Divide the batter among 8 cups of a greased patty or muffin tin—they should be about two-thirds full. Alternatively, you may line the tin with paper cases. Bake for 20 to 25 minutes.

Blueberry and Cassis Sauce

The tastes of two berries—blueberry or bilberry and blackcurrant—are brought together to grace this deep-toned sauce. It is pictured as an accompaniment to duck breasts, sliced into long strips and attractively fanned out on a plate. Serve this sauce warm over duck or any other game bird you enjoy.

(Serves 4)

50g/2 oz sugar
50ml/2 fl oz water
50ml/2 fl oz red wine vinegar
120ml/4 fl oz dry red wine
475ml/16 fl oz tinned beef stock
50–120ml/2–4 fl oz crème de cassis
150g/5 oz blueberries or bilberries
15g/½ oz unsalted butter

Put the sugar and water in a small, heavy saucepan. Cover and heat until the sugar is dissolved. Remove the lid and cook until dark, stirring frequently and being careful not to burn the sugar. Remove the pan from the heat and carefully add the vinegar (it will bubble). Add the wine and beef stock and reduce by half; add crème de cassis to taste. Keep the sauce warm.

Add the blueberries to the warm sauce just before serving. Stir in the butter and spoon the sauce over the main course.

CAPE GOOSEBERRY
Physalis peruviana

Growing in the berry patch, the Cape gooseberry has an unreal quality about it. The plump pentagonal pods that contain the berries form a natural gift box containing the berry treasure. When you crack open the thin, papery, delicately veined husks, you discover the yellow fruit within, perfect little suns. Because of its close cousins, the tomatillo and the strawberry tomato, the Cape gooseberry exists in the gardening no-man's-land between what is popularly thought of as "vegetable" and "fruit," but once you taste this oddity's citruslike tang, you will agree with the latter classification.

The straw-colored pods are actually the plant's calyx, which inflates like a balloon and colors slightly after being fertilized. It is from this trait that the entire *Physalis* genus gets its name: *Physalis* comes from the Greek for "bladder." Many other varieties of the genus, such as the Chinese lantern (*P. alkekengi*), are grown solely for the ornamental calyx, which turns to a deep orange in some species in autumn. But you will value Cape gooseberry for more than its looks.

The raw berries are delectable when eaten plain, though someone I know prefers them raw drizzled with chocolate. Some people, upon hearing the word berry, may get their expectations up for a very sweet fruit; so for such people, have honey and cream on hand when introducing them to the Cape gooseberry's subtle charms. Be careful never to eat the delicate husks, however, as they are thought to be poisonous.

The nomenclature of the Cape gooseberry is a messy business indeed, and you will see people look at them and say, "Ah, yes, a ground-cherry," or "Where did you find a strawberry tomato?" or even "It's a tomatillo, isn't it?". One native of India exclaimed upon seeing the berries, "Raspberries!" as they are known in his country. It is your duty to inform them that the first two berries are the North American plant *P. pruinosa*, whose berry is more of an orange-yellow fruit with a tomatolike taste and whose calyx is quite hairy. The tomatillo is actually *P. ixocarpa*, a Mexican native whose fruit is somewhat sticky and greenish and is used in salsa and other spicy foods. As for the Indian readership, please turn to the section on raspberries.

The origin of the Cape gooseberry is uncertain. It is a member of the nightshade family. In his book *Of Plants and People* (1985), Charles B. Heiser, Jr., writes that the species originated in the Andes, where it is an "encouraged weed" rather than a cultivated plant. Occasionally, Mr. Heiser reports, the fruits are sold in markets in Ecuador, where the plant is known as *uvilla* (a variation on the Spanish for "grape"). Other South American countries have dubbed it *uchaba* or *capulí*. Further-

more, according to Mr. Heiser, this anomalous plant was transported to the Cape Province of South Africa and entered into cultivation. (One can imagine the berries traveling very well, snugly tucked away in their compact sacs.) Thereafter, the berry trekked to New South Wales, Australia, where it took up residence under an assumed name, cultivated as the so-called Cape gooseberry.

Should you wish to grow these berries, sow the seeds in late winter or early spring in a cool greenhouse or cold frame in sandy soil. When you transfer the seedlings outdoors, plant them about one foot apart in a sunny, unobstructed and frost-free location. You can also grow Cape gooseberries in pots when the summers are cooler, as you can tomatoes.

Cape gooseberry pods look like exotic decorations in the early morning sun at this California ranch. These berries were in season in mid-October and ready to be harvested and enjoyed for their sprightly, citrusy taste.

Encased in its delicately veined pod, opposite page, the Cape gooseberry awaits its unwrapping. When its pod is unfurled, the Cape gooseberry's golden fruit is revealed, as shown on the following page.

Red and Yellow Berry Tart

This luscious dessert is not only a wonderful blend of tart lemon, tangy Cape gooseberries and sweet raspberries, but also a vivid display of the berries' rich colours. Here the Cape gooseberry complements the tart both visually and as a tasty garnish. The husks, which look so attractive surrounding the berry on the plant, also look lovely in this dish. *Eat only the delicious plant, however, since the husks are said to be poisonous.* The illustrated dish was garnished with Raspberry Berry Paint and Kiwi Berry Paint (see page 133 for a recipe for Berry Paint).

(Serves 6 to 8)

Shortbread Biscuit Crust:
175g/6 oz unsalted butter, softened
50g/2 oz icing sugar
½ teaspoon vanilla essence
175g/6 oz plain flour

Lemon Cream Filling:
4 egg yolks
100g/4 oz sugar
250ml/8 fl oz soured cream
2 teaspoons vanilla essence
grated zest of 2 lemons

Berry Topping:
675–900g/1½–2 lb raspberries or
* blackberries, boysenberries,*
* blueberries or bilberries,*
* strawberries or a combination*

Glaze:
100g/4 oz redcurrant jelly
2 tablespoons water

Garnish:
sweetened whipped cream
Cape gooseberries
Berry Paint (see page 133 for a recipe)

Shortbread Biscuit Crust: Preheat the oven to 180°C/350°F/gas mark 4. In a mixing bowl, cream the butter, sugar and the vanilla until thoroughly blended. Add the flour. Use a dough hook of an electric mixer or stir until the flour is just incorporated. With your hands, pat the dough into a fluted 23-cm/9-inch flan tin with a removable bottom until it is evenly distributed. Bake for 20 to 25 minutes, until golden brown. Allow to cool. (Leave the oven on.)

Lemon Cream Filling: In a mixing bowl, beat the egg yolks and sugar for 1 minute at medium speed. Add the soured cream, vanilla and lemon zest. Beat for 1 minute. Pour into the cooled crust and bake in the oven for 15 to 20 minutes, until the filling puffs up and is firm and golden. Allow to cool. Remove the tart from the tin.

Berry Topping: Arrange the berries attractively on the cooled filling.

Glaze: In a small saucepan over medium-high heat, bring the redcurrant jelly and water to the boil, stirring constantly until smooth. Reduce for 1 or 2 minutes. With a small pastry brush, apply the glaze to the berry topping.

Chill the tart for 1 hour, either uncovered or covered in a large plastic container that doesn't touch the tart. Garnish each serving with whipped cream, Cape gooseberries and Berry Paint.

CLOUDBERRY
Rubus chamaemorus

Judging by its evocative name and delicate color, the cloudberry would seem to be a fragile fruit, requiring tender nurture in a cloistered garden. But this is all a false front. The cloudberry is actually a rugged and solitary soul of the North, growing in marshy and peaty areas of Lapland, in the Sable Islands of Canada, and, much more rarely, in parts of Scandinavia below the Arctic Circle, and in Scotland as well.

This bog dweller at first boggled John Tradescant the Elder, the seventeenth-century gardener to England's King Charles I and avid plant hunter. Tradescant and his son, John the Younger, trekked around the world in search of rare plants to bring back by the shipload to the manor gardens of the great houses of Europe. The first Lord Salisbury was one of the Tradescants' most enthusiastic clients, commissioning expeditions to France, Holland, Belgium, and Italy to furnish the gardens at Hatfield House with fruit and flowers. In 1618, John the Elder wangled a spot aboard a ship that was departing from England on an ambassadorial mission to Russia. No Englishman had yet investigated northern Russian flora, and Tradescant was willing to endure the torturous six-week voyage in order to be the first.

As soon as the boat docked in a Russian port, Tradescant took to the countryside. He reported "many sorts of beryes, on sort lik our strawberyes but of another fation of leaf." The berry in question turned out to be *Rubus chamaemorus*. Tradescant described how the amber-peach fruits were used by the locals as medicine to prevent scurvy. He was so fascinated by these berries that he dried them "to get seede."

If Tradescant had tried to grow the cloudberry back in England, he would have been disappointed. Finns don't even try to cultivate them. One year, *lakka*, as the cloudberry is called in Finland, might emerge in the wild, while another year it may not appear at all. Even if Scandinavians, Soviets, or Sable Islanders are fortunate enough to have cloudberry growing naturally on their property, there is no guarantee that this ornery berry will compliantly fruit each year, especially in the more temperate areas of Scandinavia and the Soviet Union. Cloudberry is more reliable in the far northern regions, in or near the Arctic Circle, where the nightless days of summer can last for two months. There, showered with an inexhaustible supply of golden, growth-stimulating sunshine, the cloudberry literally glows in the boggy landscape, making it almost too pretty to pick.

The cloudberry's shape, resembling a billowy cumulus cloud, inspired its English common name. Berry world gossip hints at the imminent introduction of a cloudberry adapted to warmer climes.

But Scandinavians cannot resist the temptation. Cloudberries are a delicacy that draws long lines at outdoor markets. The yellow berries give off a musky odor—in truth, a bit strong for uninitiated noses. When eaten raw, the cloudberry's seeds tend to make the berry rather chewy. In the hands of a knowledgeable cook, the cloudberry changes character: it can be sweetened for desserts, puréed in mousses, or even made into a sweet liqueur that retains the odd smell of the cloudberry and allows its subtle flavor to emerge in the strange aftertaste. (The experience can be compared to eating *chèvre* for the first time.) The cloudberry's complex, bittersweet taste is a sharp contrast to the cloying sweetness of tropical fruits, indicating that this northern fruit is of a sterner, more Lutheran nature than most of its fairweather brethren.

This dwarf herbaceous species has a creeping stem, with erect shoots three to eight inches high. *Chamae* is Greek for both "dwarf" and "on the ground," which perfectly describes this berry's low-lying habits. It does not form the large, congenial berry colonies of other Scandinavian species like the lingonberry, but is somewhat antisocial. Its solitary white flowers resemble those of the blackberry. Botanists disagree on whether it is more closely related to the blackberry or the raspberry. Confusingly enough, it is sometimes referred to as the *dwarf mulberry*, which is how its species name, comprised of genus and species names, is literally translated.

Anyone who has experienced the cloudberry's taste knows that comparisons to other berries should not be made. It's best simply to appreciate this berry on its own terms and sample age-old Scandinavian dishes carefully developed to exploit the cloudberry's merits.

Consumption of the lakka, or cloudberry, is an ages-old tradition in Finland, and the berries are regarded as a delicacy in the open-air markets, above. Finns cheerfully line up to buy the berries, as they are difficult to come by in the wild. To the uninitiated, the cloudberry's taste can seem somewhat bitter, but Scandinavians eagerly devour the berries fresh and stirred in a bit of sugar, left.

Swedish Pancakes with Fresh Cloudberries

Here is a traditional Swedish dish that allows the unique taste of the cloudberry to assert itself. *Plättar*, or Swedish pancakes, are served as part of the traditional Thursday night meal in Sweden, followed by the main dish of pea soup with pork. As pictured here, they are served from pan to plate, enlivened by cloudberries lightly stirred in sugar.

(Serves 4; makes 8 pancakes, 6-cm/ 2½ inches in diameter)

> 2 eggs
> 2 tablespoons wholemeal flour
> 350ml/12 fl oz milk
> 1 tablespoon sugar
> pinch salt
> 225g/8 oz cloudberries
> sugar to taste
> whipped cream

Whisk all the ingredients—except the cloudberries, sugar to taste and whipped cream—in a bowl. Heat an 18-cm/7-inch frying pan and brush with butter. When the pan is hot, pour in the batter to make 8 pancakes, each about 6 cm/2½ inches in diameter. Cook over medium-high heat until golden brown on each side, about 2 minutes. Stir the cloudberries and sugar together. Serve the pancakes with the sweetened berries and the whipped cream on top.

CRANBERRY
Vaccinium macrocarpon

When the foliage on the trees begins to turn golden on Cape Cod, the cranberry harvest commences in the bogs. Harvesters hasten to gather the tart ripe fruits in the brisk Massachusetts weather. So unlike the sweet fruits of summer, the hardy cranberry belongs to the crisp chill of autumn and has been an integral part of New England life for centuries.

The Wampanoag Indians of New England told an intricate legend about how the first cranberry came into being. It seems that when the Old Marsh Woman cut her finger, she did not bleed like mortals; instead, the waters of the marsh began to pour out of her. When her brother, Maushop the Giant (who somehow had developed blood in his veins), heard the Marsh Woman's cries, he realized that her water "bleeding" couldn't be stopped. He pricked his own finger to get blood to apply to the Marsh Woman's wound. As he reached his hand toward her, a drop of his blood was blown by the wind into the cattails. The blood took hold in the soil and turned into the first cranberry.

Fanciful as the story is, there is a sound reason why cranberries grow in swampy areas. The large cranberry, *Vaccinium macrocarpon*, is a creeping evergreen vine native to North America. It is finicky about its growing conditions, requiring a low, damp location and an acid peat soil. You will find the wild bogs nestled between the dunes of the Cape Cod National Seashore, especially in the Province Lands of Provincetown.

Smaller cousins of the American cranberry flourish in wet heathlands in Britain, and are even collected and eaten by a few enthusiasts. *V. microcarpon*, the small cranberry, is native to Scotland and northern Europe. *V. oxycoccus* also grows in Scotland, but is found too in areas of England, Wales and Ireland.

Commercial cranberry bogs now abound in Cape Cod as well as in southern New Jersey and Wisconsin. When farmers cultivate cranberry bogs, they create the same conditions as those found in the wild, but even more precisely regulated. They dig ditches around the perimeter of the bogs and across them to regulate the water levels and use the ditches to flood the bog to prevent frost damage, although sometimes sprinkler systems are used.

Cranberries were called "crane berries" by the American settlers because the flower stamens form a "beak" resembling that of a crane. The Indians introduced the colonists to the berry, showing them how to use it for pemmican cakes. But the colonists soon Europeanized the cranberry, using it stewed and sweetened in puddings and tarts, for the unsugared cranberry will set your teeth on edge with its tartness.

The first cranberry commerce was begun with the pilgrims, who sent them to Europe and the West Indies. The firm berries kept very well on the long voyages. John Josselyn, in his catalog of New World flora, wrote a bit of public relations hype on the cranberry to be read by European audiences: "At first they are a pale yellow Colour, afterwards Red, and as big as a Cherry; some perfectly round, others oval, all of them hollow, of a sower astringent taste. . . ." Furthermore, he enthused, they are "excellent against Scurvy." Europeans immediately warmed to cranberries and cooked them much as they would lingonberries, in sauces, pastries, and tarts, adding plenty of sugar.

Crafty colonists also attempted to entice Europeans into buying cranberries for cultivation. American cranberries were introduced on the European continent as "a plant of easy culture, and with but little expense, meadows which are now barren wastes might be converted into profitable cranberry fields."

But this was false advertising. The cranberry is not at all easy to grow. Starting a new cranberry bog requires five years of tending before the first crop will be ready. The grower must clear the bog area of all plants, level the site, prepare the soil, plant the vines, and install a watering system. It's

Wild cranberry bogs rely on the forces of nature and the shifting of sand to produce berries. This one on the Cape Cod National Seashore at the Province Lands Visitor Center in Provincetown, Massachusetts, comes to full fruition in October.

also wise to keep several beehives so that the cranberry flowers will be well pollinated. Thereafter, the berry plants need to be pruned and the vines' runners have to be covered with sand to produce a more upright crop. Cranberries are so sensitive to weather changes that growers must stay constantly on guard. The spring growths are particularly susceptible to frost, so growers usually flood bogs to protect them. Such a regimen could hardly be classified by anyone as "easy culture."

The cranberry became even more of a boon to New Englanders when one ingenious local unlocked the key to the natural bog. Between 1813 and 1816, Henry Hall of Dennis, Massachusetts, observed chang-

ing conditions in wild bogs and discovered that more robust cranberries result when sand drifts across them. The practice became established, and the cranberry industry began to grow on Cape Cod, eventually surpassing even the fishing industry.

Developments in harvesting methods soon ensued to keep pace with the bounty of berries produced by new bog cultivation methods. In the early nineteenth century the harvesters, who were usually women, had to creep on their knees through the cranberry fields, picking the berries by hand and dropping them into their oilcloth aprons. It was tedious work, and they had to wrap strips of linen around their fingers to protect them. But then cranberry

Cranberries obligingly float to the surface to be harvested at a commercial bog on Cape Cod. The concept of the artificially created bog was developed by Henry Hall of Dennis, Massachusetts, early in the nineteenth century. Hall observed seasonal changes in wild bogs and developed a method to mimic the same conditions.

Commercial bog operators harvest cranberries using a variety of methods. This piece of equipment loosens cranberries from vines growing beneath the water.

76

scoops—boxes with wooden "fingers" to detach the berries—were developed, and this accelerated the work. Today the berries are harvested by mechanical pickers or through flooding, which loosens the berries from the vines and causes them to float.

Meanwhile, Henry David Thoreau—the American writer famous for his work *Walden, or Life in the Woods* (1854), which contained his ruminations on the natural and the human world—was not at all pleased with this development of cranberry cultivation. He was so angered, he set pen to paper and huffed that in the New England meadows near Walden Pond he "admired, though I did not gather, the cranberries, small waxen gems, pendants of the meadow grass, pearly and red, which the farmer plucks with an ugly rake, leaving the smooth meadow in a snarl, heedlessly measuring them by the bushel and the dollar only, and sells the spoils of the meads to Boston and New York; destined to be *jammed*, to satisfy the tastes of lovers of Nature there."

It's hard to be as ascetic as Thoreau. Most of us can appreciate their beauty, but we enjoy eating them, too. The recipes that follow should give you a nice overview of the cranberry's diverse uses in cooking.

Cranberry Beurre Rouge

This sauce is a delightful complement to grilled fish. A stunning way to present it is in an attractive gravy boat on a table decorated with a centrepiece of fresh red roses and cream-coloured linen.

(Serves 4 to 6)

> *100g/4 oz cranberries, fresh or frozen*
> *250ml/8 fl oz dry red wine*
> *120ml/4 fl oz fresh lemon juice*
> *1 tablespoon red wine vinegar*
> *2 tablespoons chopped shallots*
> *10 whole green (water-packed)*
>   *peppercorns, drained and rinsed of*
>   *brine*
> *2 bay leaves*
> *100g/4 oz unsalted butter, cut into*
>   *2.5-cm/1-inch chunks*
> *salt to taste*

Cook cranberries in 60ml/2 fl oz water for 3 minutes. Purée in food processor or blender. Strain through a sieve. Place the wine, cranberry purée, lemon juice, vinegar, shallots, peppercorns and bay leaves in a non-stick saucepan and boil the liquid until reduced to about 50ml/2 fl oz. Pour the liquid through a sieve and use a rubber spatula to force some of the cooked shallots through. Be sure to scrape any shallots from the underside of the sieve and add to the liquid. (This recipe can be prepared ahead of time up to this point.) Just before serving, heat the liquid and whisk the butter into the pan, a piece at a time, beating constantly. As soon as the butter has melted, remove the pan from the heat. The butter and wine reduction will form a dark rose-coloured emulsion. Add a little salt, to taste.

Cranberry Nachos

Surprising as this appetizer sounds, it is actually a delicious and inventive way to serve cranberries. The autumnal colouring of the vegetables makes the dish a perfect choice for an autumn cocktail party or even as a prelude to a harvest dinner. The medley of crisp nachos and creamy coulis and salsa makes this dish more of a feast than a finger food.

(Serves 4)

Red and Yellow Pepper Coulis
(50ml/2 fl oz each):
1 whole yellow pepper
1 whole red pepper
4 tablespoons olive oil
salt and freshly ground pepper to taste

Salsa (250ml/8 fl oz):
50g/2 oz cranberries
50ml/2 fl oz fresh orange juice
¼ medium red onion, diced
1 fresh jalapeño pepper, diced
2 tablespoons chopped cilantro or
 coriander
50ml/2 fl oz fresh lime juice
pinch salt
freshly ground pepper to taste
sugar to taste

Nachos:
12 blue and gold tortilla chips
100g/4 oz goat's cheese (preferably a
 mild Californian or French variety)
radicchio
4 sprigs cilantro or coriander

Red and Yellow Pepper Coulis: Char the whole peppers over an open flame or under a preheated grill. Place in a metal bowl and cover tightly with cling film. Allow to sit for 15 minutes, then wash off all charred skin and purée in 2 batches (1 for each pepper) in the food processor until smooth. Drizzle olive oil over each coulis batch and season with salt and pepper.

Salsa: Cook the cranberries in the orange juice for 3 minutes, until tender. Place the cranberry mixture in a bowl and add the onion, jalapeño and cilantro. Toss with lime juice and add salt, pepper and sugar to taste.

Nachos: Preheat the grill. Arrange the blue and gold chips on a baking sheet. To each chip, add 1 teaspoon each of yellow and red pepper coulis and top with salsa and goat's cheese, broken into pieces by hand. Put the pan under the grill until the cheese softens (about 5 minutes; it will become slightly browned). Remove from the grill. Arrange the chips on a serving plate. Form a bowl out of radicchio leaves and fill with extra salsa. Garnish with cilantro sprigs and serve immediately.

Cranberry Muffins

These cranberry muffins have a sophisticated flair, using such delectable flourishes as Grand Marnier and nuts. The recipe comes from Massachusetts, and is a speciality during the autumn cranberry harvest, when visitors to Cape Cod's cranberry bogs crave the tangy berries.

(Makes 8 muffins)

150g/5 oz unbleached flour
100g/4 oz sugar
2 teaspoons baking powder
¼ teaspoon salt
grated zest of 1 orange
100g/4 oz cranberries
35g/1¼ oz walnuts or pine nuts, chopped
1 egg
2 teaspoons soured cream
50ml/2 fl oz orange juice
1 tablespoon Grand Marnier

Preheat the oven to 180°C/350°F/gas mark 4. Mix the dry ingredients with the cranberries and nuts. Add the egg, soured cream, orange juice and Grand Marnier. Stir until just moist. Divide the batter among 8 cups of a greased patty or muffin tin—they should be about two-thirds full. Alternatively, you may line the tin with paper cases. Bake for 20 to 25 minutes.

Duck Liver Flan with Cranberries and Walnuts

Cranberries need not be relegated to the traditional sauce. They can be very elegant when combined imaginatively with other ingredients. This appetizer is flecked throughout with beautiful cranberry red. It is delicious when served straight from the oven. If you have any left over, chill it before serving again—preferably within two days. Warm or cold, it can be garnished with radicchio and accompanied by gherkins or a delicate sauce of mustard and crème fraîche.

(Serves 4 to 6)

450g/1 lb duck livers
9 eggs
875 ml/28 fl oz double cream
50g/2 oz chopped shallots
2 garlic cloves, finely chopped
100g/4 oz cranberries, coarsely
   chopped
1 tablespoon water-packed green
   peppercorns, drained, rinsed of
   brine and chopped
75g/3 oz chopped walnuts
2 teaspoons salt
½ teaspoon freshly grated pepper
2 tablespoons dry sherry

Preheat the oven to 160°C/325°F/gas mark 3. Process the livers in a food processor fitted with a steel blade for 30 seconds. Add the eggs and process another 10 seconds. Pour the mixture through a sieve into a mixing bowl. Add the remaining ingredients and stir well.

To prepare for baking, coat two Pyrex 18-x 7.5-cm/7-x 3-inch or 25-x 10-cm/10-x 4-inch loaf tins with a liberal amount of olive oil and line the pans with non-stick baking paper. Be sure to stir the mixture just before dividing it between the two pans. Place the pans in a water bath and bake for about 2½ hours. The flan will rise and is finished when a knife inserted in the middle comes out clean.

CURRANT

RED CURRANT
Ribes rubrum

BLACK CURRANT
Ribes nigrum

WHITE CURRANT
Ribes sativum

*W*hen early colonists left Europe for the New World, they were ready to make a new start, but they simply couldn't bear to part with certain hallowed British traditions—among them drinking tea, celebrating Guy Fawkes Day, and avidly consuming currants in homemade puddings and preserves.

The settlers were very keen on the native berries at their disposal in the New World, rhapsodizing over strawberries, blueberries, hackberries, chokeberries, and such. But nature had not been all-providing in the berry category: the robust, tangy currants that thrived in Britain weren't to be found in North America. Meals must have tasted woefully incomplete those first few lean years without this vital ingredient.

Measures were taken posthaste. Currant plants were on the list of necessities that the Massachusetts colonists ordered from London agents in 1628. But something must have gone awry, either on the ocean voyage or with the new plantings, because the next year they wrote again for "vyne plants and stones of all fruit, also wheat, rye, potatoes, barley oats, woad, saffron, lequorice seeds, hop roots and currant plants." Currant cultivation was also hindered in the South. Thomas Glover wrote glumly in his *Account of Virginia*, "Here grow good Figges and Gooseberries, but no English currants."

As time passed, the colonists learned to assuage their currant cravings by substitut-

ing other ingredients for them in traditional currant recipes. In 1671 John Josselyn, who chronicled life in the colonies in his *New England's Rarities Discovered, in Birds, Beasts, Fishes, Serpents, and Plants of That Country*, reported in a wistful tone that settlers boiled and baked bilberries (actually blueberries) for use in puddings "instead of Currence." The colonists had learned to make do with what grew around them, and as more and more frustrated gardeners abandoned currant cultivation because the British specimens did not transplant to American soil, these fruits began to disappear from their tables.

Currant connoisseurs reacted with horror. In *American Cookery* (1796), Amelia Simmons admonished gardeners for not cultivating currants. This activity "ought to be encouraged," she scolded, and as an incentive she included recipes for currant pie and currant jelly. But much of her audience—born and bred on American soil—wasn't familiar with the currant, and they continued to contentedly whip up raspberry tarts and strawberry puddings.

Meanwhile, another group of British expatriates, in New Zealand, were growing currants that would have made the first American colonists curse their horticultural fate. By the Victorian era, New Zealanders had established thriving kitchen gardens in the British mode using plants imported from their homeland and exotic native plants as well. Currants flourished in the temperate paradise of the South Pacific.

A simple, classic way to take advantage of the taste and beauty of the red currant, left, is in a summer pudding. Simply remove the crusts from slices of white bread, fit each into a fluted mold, and place crushed berries on top of them. Weigh down with a plate. The bread will become infused with crushed currant color.

Charles Darwin visited the colony in the mid-nineteenth century and recorded his impressions in his *Naturalist's Voyage Around the World*. Imagine this homesick Englishman's delight when "after having passed over so many miles of an uninhabited useless country" he encountered "a British farmhouse, and its well-dressed fields, placed there as if by an enchanter's wand." He noted the "large gardens, with every fruit and vegetable which England produces" and did not fail to mention currants in his selective listing of its contents.

What is it about the currant that has inspired such affection and loyal cultivation? Perhaps it is because the currant is so intertwined with the customs of the British Isles. But what is most inspiring about the currant, unquestionably, is its taste. *Ribes*, the plant's genus name, is the Latinized version of an Arabic word describing a plant with an acid juice. Black currants are quite tart, while reds and whites tend more toward sweetness. All are pleasantly aromatic.

Currants are indigenous to cool, moist northern regions and need to grow in places where there is plenty of summer rainfall. In some parts of the American South and Southwest it can be a challenge to grow them. The situation has been much improved since colonial times in America, however, and there are now varieties adapted to suit regions where currant production would have been an unthinkable consideration in years gone by.

Currant cultivation in the United States is not a snap, though. Plants of the *Ribes* genus, including all currants and gooseberries, are host plants for white pine blister rust, a disease that affects the eastern white pine (*Pinus strobus*). In states where this disease is a problem, gardeners are not allowed to grow these plants, so check with your local Cooperative Extension Service before planning your garden. Massachusetts, New York, West Virginia, and Wisconsin are particularly concerned about this disease, and residents of those states should be careful about currant gardening—especially when it comes to the

black currant, as this plant is a "preferred host" for such diseases.

Berry gardeners in the United States who are inspired to obtain authentic British currants should know that their task will not be much easier than that of the American colonists. Imported plants must be quarantined for two years on your property; some might not even be allowed. By all means, contact the Washington, D.C., office of the United States Department of Agriculture for information on importing berry plants of all kinds.

Such rules and regulations can make currant gardening seem complicated, but it's actually a very pleasurable activity, particularly when you've chosen the varieties that are best suited to your region. Red and white currants in cultivation today developed from the wild reds of Europe—*R. vulgare*, *rubrum*, and *petraeum*. The most highly recommended red currant variety is 'Red Lake.' Its unfussy growth requirements and medium-sized carmine berries have won gardeners' hearts on both sides of the Atlantic. 'Wilder' is another old friend, suffusing its branches with vigorous masses of berries. 'Stephens No. 9' is a prolific fruiter, and is well suited to the American Midwest.

A popular choice among the white currants (which actually have a pale yellow tinge) is the strong-growing 'White Imperial'. Also of interest, especially to the European grower, is the French hybrid 'Weisse Versailler', a Victorian-era currant with a glorious, gentle aroma.

Garden varieties of black currants are principally the offspring of the European species, *R. nigrum*. Some patriotic American garden writers contend that the wild American black currant, *R. americanum*, is just as flavorful as its Old World cousin, but in truth the European species—with its commanding, musky odor; upright, vigorous growth habit; and smooth, deep-toned berries—looms larger than all others in the currant world.

The European black currant is not without its flaws, however. When bruised, the

foliage emits an unpleasant odor, so when Alice B. Toklas, the celebrated companion of Gertrude Stein, wrote in her *Cook Book* that "the bushes of black currants have a very strong agreeable fragrance," she was probably not referring to the leaves. Gardeners should pardon this cantankerous old berry its idiosyncrasies; the rewards of a bounteous black currant harvest far outweigh its olfactory impositions. American gardeners should take pains to procure rust-resistant black currant varieties, such as 'Coronet' and 'Crusader.' Northern European growers will do well with 'Goliath', a hardy soul that doesn't mind a spot of cold weather, and 'Daniels September', which sounds like a dime-store romance novel but is actually an old English species that fruits prolifically late in the season.

The venerable currant deserves an extra bit of garden planning to do justice to its history. Some growers pamper their currants by adding a beehive to the garden to promote cross-fertilization and prodigious fruiting. Others emphasize the rustic cot-

Luminous white currants, above, are as stunning as they are delicious. These pearly currants usually produce smaller harvests than do their red and black relatives. Nevertheless, they are cherished for their taste, which is slightly less tart than other types of currants.

Black currants groan on branches in a Finnish garden, opposite page. This farmer puts the currants to use at home by making them into a rich maroon-colored currant-strawberry drink, which he recommends consuming with smoked fish.

tage-garden quality of the plant, creating a small "grove" of currant shrubs with a simple old-fashioned wooden bench nearby—the perfect enclave for contemplation and conversation. On a more practical note, currants are also wonderful plants for masking an unattractive garage or shed. Simply prop a trellis against the offending structure and encourage currant plants to grow cordon-style along it.

Currants will reward you year-round for your gardening trouble. These obliging plants open their delicate pinkish-red and green flowers in early spring; berries adorn the bush in late summer and continue on into early autumn.

Some species of currants are grown for their good looks alone. The buffalo currant (*R. aureum*), flowering currant (*R. sanguineum*), and Missouri currant (*R. odoratum*) are all charmers, with graceful carriage and stunning blooms. All this vigor raises expectations for equally marvelous berries, but they are, unfortunately, quite bland in taste.

The ornamental currants were favorite plants of Vita Sackville-West, the novelist and garden writer. In her *Garden Book*, she gave the nod to the old flowering currant, a standard in the British cottage garden, clipped into hedge form. "A most reliable shrub," she wrote, "never taking a year off, and demanding the minimum of care and cultivation. . . ." She enthused, too, over the buffalo currant for its lovely April-blooming yellow flowers, which "have the advantage, for those who like cloves, of diffusing that spicy scent, and there is further advantage that the leaves in autumn will turn to a fine gold."

Certainly these attributes are more than enough to justify growing the ornamental currants, but what berry gardener hasn't dreamed of how lovely it would be to combine the flowers and fruits of the edible and ornamental varieties to produce the perfect currant? For complicated botanical reasons, this doesn't seem destined to happen, so you'll have to plant both kinds for the full show of beauty and berries.

Of course, edible currants are not without their own beauty. Red currants are colorful accents in fruit salads and add piquant flavor. They also are lovely in a summer pudding, which entails pressing bread into a mold, adding currants, and letting the concoction set, as the berries impart their gorgeous ruby color and bittersweet taste. You might also heap a dish with clusters of different-colored currants—perhaps with a few red raspberries scattered throughout—to create a centerpiece reminiscent of a Dutch school still-life painting.

Most of all, currants are esteemed for their culinary versatility: they can be used in summer soups and make succulent sauces for chicken, pork, and duck. They are also outstanding as sorbets and preserves. Black currants are often used for wine-making, a homespun hobby pursued by many a respectable berry grower in England. In France, the black currant liqueur, *crème de cassis*, is the principal ingredient, along with white wine, in a drink invented by Canon Kir, the Mayor of Dijon in the 1950s. This drink is particularly refreshing in summer with a bit of shaved ice. The French also enjoy the black currant, or *cassis*, in sherbets and compotes, as a syrup, and in preserves and candies. Finns prefer vitamin-packed black currant—strawberry drinks to get them through the frigid winters.

American perseverance in attempting to grow the currant has finally resulted in a fair though spotty representation in the American garden. These beautiful red currants, opposite, grow in Croton-on-Hudson, New York. The European black currant, left, a Eurasian native, will produce bumper crops if you plant several varieties to ensure cross-pollination.

Ambrosial Berry Salad

Fresh berries deserve equally vibrant ingredients to accompany them in dishes. This sprightly salad of three berries also contains fresh greens and creamy goat's cheese. A splash of blackcurrant vinaigrette heightens the taste of this delightful mix. The beauty of the dish lies in its simplicity.

(Serves 6)

Blackcurrant Vinaigrette (250ml/8 fl oz):
175ml/6 fl oz blackcurrant vinegar
3 tablespoons extra-virgin olive oil
2 teaspoons freshly ground pepper
1 teaspoon finely chopped shallots
½ teaspoon finely chopped garlic
50g/2 oz fresh or defrosted frozen
  blackcurrants
salt to taste
sugar to taste (optional)

Salad:
12 leaves Belgian endive
3 half-heads of seasonal lettuces (other
  fresh seasonal greens can also be
  used)
225g/8 oz blackberries
225g/8 oz golden raspberries
225g/8 oz blackcurrants
12x25g/1 oz slices mild goat's cheese
bachelor's buttons (any edible flower
  can be used)

Vinaigrette: Mix the vinegar, olive oil, pepper, shallots and garlic well. Add the whole blackcurrants and salt to taste. Vinaigrette should taste tart, but if you prefer a sweeter vinaigrette, add a small amount of sugar to taste. Cover the mixture, refrigerate it and allow it to sit overnight for the flavours to marry.

Salad: Toss the vinaigrette with the salad greens, and divide equally between plates. Distribute the blackberries, golden raspberries and blackcurrants among the plates. Add 2x25-g/1-oz slices of goat's cheese to each serving. Garnish with colourful bachelor's buttons.

If currant cultivation isn't your fancy, or if you are a United States resident who can't grow _Ribes_ species plants in your area because of rules pertaining to the spread of white pine blister rust, try to locate a farmers' market that sells them so that you can obtain them fresh, facing page.

ELDERBERRY
Sambucus nigra

The elder, as its name suggests, commands the respect reserved for the aged. This showy shrub, a familiar sight in woodlands and hedgerows, puts forth floating white masses of flowers surrounded by feathery leaves in midsummer and follows with enticing, pendulous bunches of red or black berries in late summer. Since olden days, countryfolk have eagerly awaited the elder's seasonal bounty and made use of virtually every part of the shrub, feasting on the flowers, using the root juices as a tonic (not recommended), and imbibing the berries in wine.

With such a prodigious harvest to be reaped from a single plant, it's no wonder that the elder is steeped in folklore. The plant's genus name, *Sambucus*, refers to the ancient Greek word for a musical wind instrument, *sambuce*. The woody stems have a large central pith that can be removed, leaving a hollow space in the plant that serves admirably as a flute or whistle.

A plant that can be transformed into an instrument seems enchanted, indeed, and many powers are attributed to the elder. It has long been considered a protective force against evil. British folklore has it that elder wood can undo a sorcerer's spells. In Denmark, rustics have believed for centuries that a creature called the Elder Mother watches over the plant, so no one dared harm it. The elder has inspired such respect, in fact, that there is an old European custom of tipping one's hat to the elder tree as a measure of respect.

The elder is also a Christian motif. It is known as the "tree of sorrow" because of the belief that Christ's cross was made from the wood of the tree. This is why Gypsies will not use elder wood in a fire and in olden days hedge trimmers were afraid to cut it. A common European practice of yore was to prune an elder bush in the shape of a cross and plant it on a grave. It has long been associated with the dead: the shrub's wood was once burned with corpses, presumably to ensure safe entry for passengers to the afterworld.

The lore of the elder even extends into medicine, with a strong dose of magic added. There is an old Bohemian custom of brewing a medicinal tea or wine from elder flowers, but the beverage will not have any value unless the flowers are gathered on Midsummer Eve (June 23). In Britain, the elder was also once considered a remedy for warts: the charm was to hang a snail on the elder shrub, and as the creature wasted away, so would the warts. In Victorian times, the elder evolved into a cosmetic aid. Elder-flower water was a staple of every cultured woman's toilette, for it was considered efficacious in the removal of freckles and dreaded suntans.

The elder is a striking plant and sometimes grows to as high as thirty feet, at which point it is considered a small tree rather than a shrub. Its sprawling form makes it an excellent choice as an informal planting, and it will lend a rustic touch to its surroundings. It was planted prolifically in England—against walls, in gardens, in

churchyards, and anywhere that its spreading form would add a gracious design note and perhaps a bit of good luck. The garden writer Gertrude Jekyll took note of the elder's popular presence in Britain, but she appreciated it for reasons other than its purported magical powers. "I am very fond of the Elder-tree," she wrote. "It is a sociable sort of thing; it seems to like to grow near human habitations. In my own mind it is certainly the tree most closely associated with the pretty old cottage and farm architecture of my part of the country."

In Britain, elder plants were once cultivated in orchards for wine-making. Although this practice has all but died out, elderberry wine is a delicacy that should be enjoyed more widely. It is now made at home in locales as diverse as Appalachia and Britain. Miss Jekyll also enthused about elderberries, recommending that they be "made into Elder wine, to be drunk hot in water." Elderberries are unpalatable raw; they must be cooked before they are used and are lovely in pies and preserves. The berries are rich in vitamin C, so you will be able to derive medicinal benefits—rather than just delightful taste—from your culinary efforts.

Fresh elderberries are the delight of birds, but not of people. Unless elderberries are cooked, their taste is insipid. Look for the elderberry's showy, lacy, flat-topped white flowers around midsummer, so that you can return later to forage for the fruits.

GOOSEBERRY
Ribes Grossularia

The gooseberry has had to struggle for recognition. It is believed that the Normans brought it into France from Scandinavia in the tenth century, but this plump, flavorful berry failed to inspire gardening enthusiasm there, earning the elaborate but ungainly name *groseille à maquereau* because the berries were principally cooked with mackerel. The gooseberry made so little stir in France, in fact, that the British Isles didn't encounter it until several centuries later, and it wasn't widely cultivated in England until the early sixteenth century.

A good part of the gooseberry's reputation seems to be determined by its thorns. While eating gooseberries is a joy, harvesting them can be quite a prickly experience. In fact, "Old Gooseberry" is a nickname for Satan, a natural extension of the moniker "Old Scratch," the connection, of course, being the spines. To quote *Webster's Dictionary*, Old Gooseberry is "the personification of evil," a terrible burden to place on this most amiable berry. The image has been perpetuated through remarks such as Vita Sackville-West's scathing simile in her *Garden Book*: "as cross and spiny as a gooseberry." To make matters worse, *gooseberry* is also slang for a chaperon, the idea being that the escort sticks to the young lovers like a thorn. (Ever resourceful, British countryfolk found a way to put the real gooseberry to good use:

Gooseberries favor northern regions, particularly England, where the cool climate and generous rainfall make it possible for them to thrive. Too much summer sun is anathema to this berry.

an old remedy for wart removal is to pierce it with a gooseberry thorn passed through a golden wedding ring.)

And then there's the matter of the gooseberry's frivolous common name, which makes it difficult to take the berry seriously. The name probably derives from the culinary practice of serving the berry with goose, either stuffed inside or served as a sauce, although there are alternative linguistic theories too involved to delineate. In parts of Britain, you might also hear it referred to as *goosegog*.

Despite some of the indignities it has suffered, however, the gooseberry has been enjoyed in Europe, principally Britain, for centuries—a testament to its tart taste. There is an ancient and elaborate German ritual that illustrates the enthusiasm some people have for the gooseberry. On Midsummer Day (June 24) children in the Eifel Mountains used to fashion wreaths and posies from various wild plants and cast them on rooftops. There they remained until the wild gooseberries were ripe, when the children would retrieve the plant decorations, set them on fire, and run with the burning plants in hand to the wild gooseberry bushes to fumigate them—presumably ridding them of evil spirits. Only after this rather dangerous ritual took place could the gooseberries be safely picked and eaten by the children. Quite a bit to go through for fresh gooseberries, but well worth the trouble to the mountain people.

The Englishwoman Mrs. Isabella Beeton extolled the gooseberry's virtues in the nineteenth-century work *The Book of Household Management*: "All sorts of gooseberries are agreeable when stewed, and, in this country especially, there is no fruit so universally in favour."

While this is still arguably the case in England, the same cannot be said of gooseberries in America. The New Englander who wrote in 1634 that the wild colonial gooseberries were "little inferior to those that our grocers sell in England" was being kind to the American gooseberries. Gooseberry gourmands say that the American fruit cannot rival *R. Grossularia*, the garden gooseberry of Europe. In her eponymous cookbook, Alice B. Toklas was more to the point and brutally honest: "The gooseberries in France are four or five times larger than those grown in the United States, and very much sweeter."

The European garden gooseberry, a native of Eurasia and northern Africa, grows three to four feet high and bears a lovely fruit. The berries are generally a bit bristly and usually greenish but sometimes also yellow or even red. This shrub, also known as the English gooseberry, does not grow easily in North America and will thrive only in northern areas and near the Great Lakes, simply because the sun is much too strong in other parts of America.

Unfortunately, the gooseberry shares the currant's unpleasant penchant for hosting white pine blister rust, so if you are a United States resident, you must contact your Cooperative Extension Service for details on whether you are permitted to grow gooseberries in your area.

Green varieties of European gooseberries, right, are valued for jam making as well as for eating fresh.

HUCKLEBERRY
Gaylussacia spp.

There is a certain mystique to the huckleberry. Wily yet beloved, the huckleberry balks at growing in the home garden, preferring to rough it in the wild, which may have inspired Mark Twain to name his famous character Huckleberry Finn after it. Huckleberry hunting might not suit you if you are a delicate type. Since the plants thrive at high elevations and deep in the woods of North America, a quest for these berries will most likely involve a lot of climbing and searching. If you are fortunate enough to discover a huckleberry patch, gather buckets of berries, for the huckleberry, which ranges in color from jet black to maroon, is a great favorite of bears and other wildlife.

It's also questionable whether you will find the berries in the same remote location the next season. Although huckleberries are notoriously unreliable, they are likely to be found in burned-out areas where the ground has been nourished by a fire's ashes. As with blueberries, distant cousins that are also members of the heath family, huckleberries thrive in acidic soil.

Despite the many hindrances to finding huckleberries, lovers of this fruit forgive all when they actually get around to eating it. The taste has been variously described as sweetly exhilarating, rambunctious and untamed, and tart and invigorating. Huckleberry hounds will also tell you that this berry is far more delectable than any blueberry—wild or cultivated. Thoreau described them as "ambrosial" but warned that the only way to experience their taste unsullied was to devour them fresh, right off the bush.

Because of the huckleberry's clandestine habits, the exact number of species is unknown and controversy rages over the botanical differences between the huckleberry and the blueberry, though some people might hazard to say that there is no difference at all. Seemingly unanswerable questions arise regarding the huckleberry's true identity: Is it a member of the *Vaccinium* or the *Gaylussacia* genus, or both? What are the differences between eastern and western North American species? To exacerbate the confusion, there is even a type of plant called the garden huckleberry, *Solanum melanocerasum*, also sometimes known as *S. nigrum*. It is a member of the nightshade family, and the leaves are purported to be poisonous, as are the unripe berries of some wild forms. Therefore, it's probably best to obtain this "huckleberry" only from a reputable garden supplier.

Occasionally some brave soul steps forward and provides hard-and-fast guidelines regarding the huckleberry. In his classic forager's tome, *Stalking the Wild Asparagus*, Euell Gibbons wrote, "If you really want to know whether the berries

Wild huckleberries grow on a bluff overlooking the Pacific on a site that a fire had recently laid bare.

94

you are picking are *Gaylussacia* or *Vaccinium*, examine the seeds. *Gaylussacia* has ten hard, seedlike nuts inside, while the *Vaccinium* has many softer seeds."

To many rugged Americans of decades past, searching for the huckleberry was not merely a pastime. By the nineteenth century, huckleberry huckstering had become a profitable enterprise in America—so much so that even Thoreau, who delighted in taking a dig at anyone who profited from nature, "thought often and seriously of picking huckleberries." The huckleberry even inspired a lifestyle, that curiosity of Americana known as *huckleberry camp*, which reached its peak in the Depression era. Families from New York State to Montana would head for the mountains in summer to forage for berries, and they canned them on the spot at their makeshift households. Camp would last until the huckleberry season waned.

Any recipe using blueberries will work with huckleberries—some would say it will work even better. Your options are limitless: pies, muffins, syrups, preserves, even wine if you're adventurous.

Warm Quail and Huckleberry Salad

The quail is widely considered to be the most distinguished and flavoursome of the game birds, while the huckleberry is revered throughout the American West as the top prize of the forager. These two uncommon foods are married in this unusual recipe. If huckleberries are not available try substituting bilberries or blueberries.

(Serves 4)

Vinaigrette (350ml/12 fl oz):
50g/2 oz huckleberries
120ml/4 fl oz white vinegar
1 tablespoon sugar
50ml/2 fl oz hazelnut oil
175ml/6 fl oz vegetable oil
salt and freshly ground white pepper
  to taste

Salad:
225g/8 oz watercress
12 leaves Belgian endive
½ head red-leaf lettuce
½ head cos lettuce
8 orange segments
25g/1 oz sliced hazelnuts
4 quails, boned and butterflied
salt and freshly ground pepper to taste
8 tablespoons clarified butter

Vinaigrette: Steep the huckleberries in the white vinegar and sugar for 24 hours. Mix thoroughly with the remaining vinaigrette ingredients.

Salad: Wash the salad greens and garnishes and arrange on 4 plates. Season the quails with salt and pepper and sauté individually in a large frying pan in clarified butter. Arrange the quails on the salads and douse with vinaigrette.

JAPANESE WINEBERRY
Rubus phoenicolasius

Some berries cling to their native habitat, fussily growing only under very specific conditions. Others, such as the Japanese wineberry, are more gracious and cosmopolitan, quietly adapting to the situation at hand. This nomadic *Rubus* genus plant exudes a mysterious Oriental exoticism no matter where it takes root.

Taste a Japanese wineberry and your thoughts might travel to the Far East: a musky, provocative quality and air of mystery characterize this succulent delicacy. Prejudice against the wineberry's taste is evident in some botanical tomes, but this is a result of the understated sweetness of the fruit. When people regard the pink-toned, luminous fruits, they prepare their palates for the honeyed sweetness of a raspberry. But the wineberry's flavor is more complex, more haunting. Sugaring the fruits slightly will bring them in line with more well-known *Rubus* berries.

The wineberry's travels through history are as curious as its taste. It is known that it is native to Japan and China, but there is some question as to when it left the East and appeared in Europe and America. In the United States, it escaped from cultivation into the wild and is known to grow prolifically in northeastern parts of the country. It has also been spotted in southern Appalachian regions, where residents sometimes call it *strawberry-raspberry* and transform it into beautiful coral-colored condiments, like jams and jellies.

When left to its own devices, the Japanese wineberry will form dense thickets, its canes arching over and rooting themselves at the tip. The canes themselves are downy with orange-reddish hairs and sport a few prickles, but nothing quite so tenacious as those of its relative the blackberry.

Aesthetically, the wineberry is a boon to the berry garden. Its leaves have beautiful purple veining and are dusted on their underside with a white felt. The tiny pink or white flowers sport a bristly red calyx. These give way to the pleasingly round berries, which literally shine and appear to be illuminated from within.

Because of these valuable ornamental qualities, the Japanese wineberry is a favorite in English gardens. The plant appears in profusion at Barnsley House, the home of garden designer and writer Rosemary Verey, near Cirencester, in Gloucestershire. Mrs. Verey, whose *potager*, or vegetable garden, at Barnsley is an inspiration to any berry gardener, emphasizes the weeping, romantic forms of Japanese wineberries. They are most striking trained along a fence in front of an eighteenth-century temple and a placid pond, themselves motifs of Oriental gardens that harmonize effectively and nicely with the wineberry plants.

In terms of recipe usage, the wineberry is to the raspberry as the huckleberry is to the blueberry. In other words, you may substitute the berry in any raspberry recipe, but allow for the subtle difference in taste to emerge. The name *wineberry* would seem to suggest that wine is made from the berry, but this is actually not the case. The name was most likely applied to the berry because of its complex taste and fragrant bouquet, vintage year unknown but no doubt quite ancient.

At Barnsley House in England, Rosemary Verey's plantings of the Japanese wineberry are stunning accents to an eighteenth-century temple situated by the side of a garden pond.

Even the form of the wineberry, right, speaks of its Oriental origins. The bristly, spidery calyces and the curved prickles resemble dramatic elements of a Japanese ikebana floral design.

The Japanese wineberry is such a beautiful landscape accent, above, that you may forget it serves another function: as a flavorful table treat.

It is tempting to leave the Japanese wineberry to its own devices and let it fruit prolifically, right. However, this berry must be kept under control periodically through pruning, because the canes take root at the tips, spreading the plant so broadly that you might soon have a wineberry jungle on your hands.

100

LINGONBERRY
Vaccinium vitis-idaea

When cool breezes of late summer ruffle the leaves of Scandinavian forests, the lingonberry season is in full splendor. Amid mushrooms of every size, shape, and hue, the forager will find the wild berries thickly covering the ground, glittering in the sporadic play of forest sunlight. Lingonberries grow so abundantly in Scandinavia and other northern climes that they are not cultivated by farmers, but simply reaped from the wild.

As is often the case with berries, particularly those that thrive in the wild, there is much nomenclature to muddle through when identifying the lingonberry. It has been called *cowberry* and *red whortleberry* (its current British names), *northern mountain cranberry*, and *foxberry*. To clear up any confusion, this species is native to northern areas of Eurasia and North America, although it is most prolific in Scandinavia and the Soviet Union. A relative of the cranberry, it is a creeping evergreen plant and has attractive glossy oval leaves. In springtime, the lingonberry decorates the forest floor with delicate, nodding pink urn-shaped flowers. Its species name translates as "the grape of Mount Ida" (the same mountain in Greece for which the red raspberry, *Rubus idaeus*, is named), which certainly sounds intriguing and romantic.

Love of the lingonberry lures Scandinavians into woodlands to gather them by the pailful. They are so plentiful, it is an easy task to gather enough for eating fresh and using in preserves in just one day.

In Scandinavia, the lingonberry is as common and sought after a fruit as the strawberry is in the United States or the currant is in England. Every effort is made to reap the wonderful berries from the wild, and berry picking there is an everyday activity. This is why, carefree as berry picking may seem, it is taken so seriously in Scandinavia—to the extent that the berry-picking theme has found its way into folklore and ancient rituals.

Above all, the purpose of such rituals was to prevent harm to innocent berriers in the forest. One ancient Finnish belief involving berries called for bear worship, but with an ulterior motive. In the ritual, a slain bear was ornamented in the hope of placating all bears who might harm berry pickers. Worshipers would cover the bear's eyes with birch bark or silver or copper coins and attach a tin plate over its nose. Other ornamentation included a multicolored cloth spread on the animal's back and sometimes even a hat and scarf on the head of a male bear and pearls and a shawl for a female. With these elaborate props in place, women and children bedizened the bear's claws with brass rings. This was done so that bears would not frighten them when they went berrying in the forest.

Another possible threat to berrying in Scandinavia and Russia is the Forest Spirit.

If you're in Scandinavia, but not in the mood for berry picking, obtain them fresh at the local open-air market, where vendors vie with each other to display immense vats of woodland fare.

This bizarre being, who alters his appearance according to his whims, lives in the forest, where he keeps a vast treasure trove. He moves from place to place in the forest—sometimes riding at full gallop on a horse, other times raging through as a whirlwind—with the intention of frightening berry pickers. He also sometimes ventures out of the forest and into the villages, where he could very well try to steal a lingonberry dish, being covetous of anything removed from his domain.

Russian folklore, too, holds berry picking in the highest esteem. A Russian "death song" documenting the travels of the departed in the realm of the dead, asks, "Did they escort thy soul over lands rich with berries, over highly beautiful heaths? Could thou with thine own hands pluck the berries?" The reference to heaths logically points to none other than the ericaceous lingonberry. The song goes on to equate the number of berries available to the departed in the afterlife with the goodness of deeds done in life. The more berries available, the less selfish the individual in life.

Symbolism aside, the lingonberry is a significant berry in the life of northern peoples simply because it is delicious. It is not really a sweet berry, but not as tart as its relative the cranberry. A favorite Finnish dish is *pronkäryistys*, or sautéed reindeer, which is often garnished with lingonberries. Lingonberry preserves are also a common Scandinavian food—a welcome treat in the dark depths of winter, the fresh taste capturing the summer rays of the nightless day. A traditional Swedish dish is called *fyllda strutar*, or pastry cones filled with lingonberries and whipped cream. It is traditionally served by placing a drinking glass in the center of a shallow glass bowl and placing a cone in the glass. Other cones are leaned up against the glass, and more cones are leaned against these until the bowl is filled.

Lingonberries usually grow among exotically colored mushrooms in many of Scandinavia's forests.

Lingonberry Sauce

Here is a wonderful autumn dish that goes nicely with marrow and wild rice. If you can't find any lingonberries, try using red currants instead.

(Serves 4)

50g/2 oz fresh lingonberries or
lingonberry preserves*
50ml/2 fl oz lingonberry liqueur
(optional)
350ml/12 fl oz rich poultry stock (duck
or chicken)
salt and white pepper to taste
50g/2 oz butter, cut into pieces

Mix all the ingredients, except for the butter, in a small metal pan. Cook on medium heat for about 3 minutes or until the sauce reaches the proper consistency—when it achieves a medium thickness and coats the back of a spoon. Before serving, swirl in the butter.

*If you use lingonberry preserves, add about 25g/1 oz of fresh berries for garnish. Redcurrants are used in that way in the photograph.

LOGANBERRY
Rubus loganobaccus

While most Victorian-era plant breeders were fussing with *Fragaria*, the strawberry, Judge James H. Logan, who was especially fond of bramble berries, let his thoughts drift into thornier pastures. In the warmth of his backyard berry garden in Santa Cruz, California, the visionary Judge Logan dreamed of introducing an entirely new plant to the *Rubus* genus.

The story goes that he began diligently experimenting and in 1881 raised a new berry from seed. It was introduced commercially the following year. The new fruit was a red-berried, upright-growing bramble whose parentage is still open to argument. Some botanists believe it is a hybrid of the western dewberry and the red raspberry, while others maintain that it is a variety of the dewberry. The difference, for those who like berry esoterica, is that a hybrid is the result of a cross between two species or subspecies, while a variety is a plant within a species that differs from the norm in some way. Some loganberry fanciers suspect that the native California blackberry *R. vitifolius* is somehow involved in the parentage.

No one is claiming that Judge Logan was an impostor, but the dispute over the loganberry's parentage has raised questions as to whether he truly invented the fruit or simply found it. After all, doubters say, the octogenarian lived until 1928, forty-six years after his berry debuted publicly, and never produced a universally acceptable parental profile of his self-named berry child. Of course, the differences among members of the *Rubus* genus are confusing to even the most learned botanists, so it's understandable that Judge Logan could have misidentified his subjects. Therefore, we'll assume that he did create the berry.

The loganberry, a vigorous grower, became a favorite soon after it was introduced. It seems that Judge Logan wasn't too concerned about easterners being able to grow his berry, because it has a limited range. This berry is perfectly suited to the Pacific Northwest and California coast, although it can be grown outside this area if heroic efforts are made to protect the trailing, prickly fruit from winter kill—usually by burying the canes in burlap and straw in cold weather. A thornless variety has a wider range and is often seen in British gardens. Here they are trained picturesquely along a wall and sometimes grown commercially on trellises.

Judge Logan must have been inspired when he invented the plant, as it puts on quite a display in the garden. The charming white flowers are followed by pendulous groups of large conical fruits. These alluring berries ripen in August on peacock-blue stalks reminiscent of grapevines when they are trained in meandering loops on trellises. They have a distinctive, slightly acidic taste agreeable to all palates and should be used as you would a blackberry or a raspberry—whichever you judge to be the dominant parent.

The loganberry is among the most handsome of the brambles, sending forth robust, elongated tangy red fruits, which reach their peak in August in most regions.

There are several varieties of the loganberry, each with its own merits. On the whole, they thrive in the Pacific Northwest and central California coast. Loganberries are also common in England, where the cool, moist climate promotes vigorous growth.

Because the loganberry has such long canes, which are inclined to trail, it is best to train the plant on wires and posts. The congenial climate of Sonoma County, California, allows these berries to continue fruiting into October.

MULBERRY
Morus nigra

The mulberry is not a bush, as the nursery song goes, but a tree—and a beautiful, stately one at that. Mulberries are known to weather centuries in good health, and they stand statuesquely as living antiques in British estate gardens. The tree, actually an Asiatic native that tends to grow no taller than fifty feet, was introduced to Britain by the Romans during the occupation.

In the sixteenth century, James I took a special interest in the mulberry, hearing that silkworms fed on the leaves. Inspired by dreams of a British silk industry rather than berry consumption, he supervised the setup of extensive mulberry orchards. It is believed that he obtained the black mulberry (*Morus nigra*), which is excellent for the table but not a silk producer, as is the white mulberry, *M. alba*. Whichever species he grew, silk never became a British export. Nevertheless, James's monetary motives led to the happy consequences of the mulberry becoming a popular and widely planted tree in Britain.

There is also a native American mulberry, the red mulberry (*M. rubra*), which in times past was often planted as a shade tree and also to provide fruit for hogs and chickens. This could lead to interesting consequences, as the unripe fruits are rumored to be hallucinogens. Its fruits are carmine, purple, and occasionally black. The black mulberry's fruits are darker as a rule. The fruits of the white mulberry, as might be expected, are white or sometimes pinkish and are considered sweeter than those of other mulberry species. All mulberry fruits—which are long, spindlelike, and resemble blackberries—are edible. Different species are preferred for various dishes, however. Black mulberries are favored for desserts; reds are nice in preserves and wines; and white are lovely dried, and used as a substitute in recipes for figs or raisins.

Of course, being such a widespread and beautiful tree, the mulberry has a rich lore associated with it. It extends as far back as Ovid's *Metamorphoses*, which relates a bittersweet story involving the mulberry. Reminiscent of Romeo and Juliet's tragic tale, the legend culminates with each of two illicit lovers mistakenly believing the other is dead and with the suicide blood of one forever staining the mulberry bark red.

In its native Japan, the mulberry was often planted as a charm to prevent lightning from striking. In China, however, it is believed that planting it in front of your house could bring misfortune. Strangely, the mulberry also is an evil fruit in ancient German beliefs. Mothers used to sing a song warning their children against eating the mulberry. It was believed that the devil used the mulberry to blacken his boots.

The mulberry has regal associations. Devotees of the tree include Queen Elizabeth I, who is said to have planted four of the trees herself at Hatfield House, Hertfordshire (although these trees are also attributed to James I). Even Shakespeare is associated with the mulberry. Historians claim that he planted one in his garden at New Place, Stratford-on-Avon. It was supposedly chopped down by the next resi-

The fruits of the black mulberry go through several stages in the ripening process — with color changing from white to pinkish red to purple to black. Don't be surprised if you find yourself sharing the berries of this Asiatic native with birds, which may even nest in its boughs.

dent of the house, the Reverend Gastrell, because he was tired of tourists tramping over his property to look at it. Fallacious remnants of "Shakespeare's mulberry tree" have been circulating around Britain in snuffboxes for centuries.

When the American colonist Captain John Smith reported on New World flora, he was struck by the contrast between the civilized mulberry and the untamed natives, writing that he spotted "great mulberry trees by the dwelling of some savages."

It seems that cultivating the mulberry simply to enjoy its fruits is a relatively recent practice. During the Middle Ages, mulberry was used in England as a coloring agent for a souplike "dessert" dubbed *murrey*, which actually means mulberry colored (gray or dark purple). In the Elizabethan era, fleshy black mulberry roots were boiled in vinegar and rose water. This dubious-sounding concoction was believed to ease toothache.

The black mulberry served yet another dental function: a powder made from its roots, its seeds, and vinegar was used to hold teeth in place. People shied away from the fruit because it could stain much-coveted porcelain-pale skin. The berries were put to use when unripe, however, as a remedy for sore throats.

As time passed, people inevitably began experimenting with the berries, perhaps taking their cue from the flocks of birds that mulberry trees inevitably draw. By the eighteenth century, mulberry wine had become a common household beverage, and mulberries were enjoyed as a dessert.

Today we have numerous options for mulberries were enjoyed as a dessert. enjoy mulberries, of course, is to eat them fresh. They are quite sweet. They can also be made into pies, jams, and jellies. A mulberry-minded resident of Somerset, England, recommends simmering them slowly in their own juices and adding a pinch of sugar to taste. Let this mixture cool and accompany with thick cream or serve with sorbet and cream. The Edwardians were very fond of this dish, she assures.

Mulberries have traditionally been planted to provide food for farm animals, and this sheep, above, finds food and shelter from the late-summer sun under the mulberry tree's leafy canopy.

Beyond the walls of Glebe Court, Lady Hobhouse's private residence in England, an ancient mulberry tree still continues to bear fruit, right. Though the tree toppled years ago, enough of the root system remains firmly embedded in the ground for the tenacious old tree to continue living.

RASPBERRY

RED RASPBERRY
Rubus idaeus and *Rubus strigosus*

PURPLE-FLOWERING RASPBERRY
Rubus odoratus

BLACKCAP RASPBERRY
Rubus occidentalis

GOLDEN RASPBERRY
Rubus idaeus

*I*f the blackberry is the fallen angel of the *Rubus* genus, then the raspberry is the pillar of bramble society. Somehow, centuries ago, the raspberry managed to disentangle itself from the dark tales of the brambles. Its reputation has emerged in the twentieth century with nary a scratch. Unencumbered by evil associations and steeped in laudatory lore, the raspberry is cultivated enthusiastically around the world. Part of the reason for the raspberry's impeccable reputation could be that it is a most amiable berry, ever ready to please with a diversity of colors and a repertoire of subtle tastes.

The raspberry even upstaged the blackberry when the genus was named: *Rubus*, the old Latin name for bramble berries, refers to the color red. *Idaeus*, the species name of the classic cultivated red raspberry, alludes to lofty Mount Ida in Greece, where, legend has it, the plant originated. Pliny classified the raspberry as a medicinal plant. The plant's high vitamin A and potassium content certainly supports this point of view. In his famous work *The Herball, or generall historie of plantes* (1597), John Gerard speaks of the "Raspie bush or Hinde berry" as having numerous "vertues," such as being an excellent balm for wounds. Indeed, the raspberry was once something of a wonder drug—which, historically speaking, means that it didn't actually harm its patients. Gerard also found it useful for holding teeth in place and for healing eyes that "hang out."

Just when you think that berry season is over, everbearing red raspberries fruit brilliantly in autumn. To ensure healthy growth and an ample supply of fruit, plant raspberries in a sunny location.

Potawatomi Indians in North America used to simmer the root of the red raspberry and then apply the liquid as a curative eyewash; this later became commonplace among American settlers. The Flambeau Ojibwa Indians used the red raspberry itself as a seasoner for medicine, while the Menominee favored the black raspberry's root, mixed with Saint-John's-wort (*Hypericum ascyron*), to treat consumption.

The raspberry is also believed to perform more spiritual functions. If you dream of the raspberry, it's said, you will enter a very pleasant love affair. There is, however, one stipulation: you must not dream of these soothsaying berries when they are out of season; this is considered unlucky for all fruits and flowers. As the old Sussex maxim goes, "Fruit out of season/Sounds out of reason." Luckily for dreamers, the raspberry is often in season, bearing a summer and an autumn crop.

Perhaps the only less than tasteful blow to the raspberry's reputation was struck in the political arena, where things so often turn sour. In the 1840 presidential campaign in America, the Whig party attacked opposing candidate Martin Van Buren for overindulging in raspberries. To do such a thing must have been considered in very bad taste; Whig William Henry Harrison resoundingly defeated Van Buren in the presidential election.

Of all the raspberries, the red raspberry is the best known and best loved. This cherished plant has arching canes that can extend as long as six feet, and sometimes they have a slight whitish tinge. An obliging shrub, the red raspberry doesn't even have as many prickles as other *Rubus* species, such as the blackberry. In the wild, it can be found growing on rocky hillsides and also in clearings. Even its leaves point to its sweet nature: they are heart-shaped at the base and have a whitened, downy underside. Raspberry leaves were brewed as tea by American colonists; the drink was known as *Hyperion tea*, named for the Grecian father of the sun god, for raspberries thrive on sun.

A pleasant morning's picking of red raspberries provides many possibilities for light meals, above. Try them in a creamy blender drink, in a fruit salad, or in a sauce to accompany poultry. The blackcap raspberry, opposite page, is valued for its use in jelly and jam making.

The red raspberry's flowers are a lovely white and blossom from May to July. The fruits, growing in fragrant bouquets, ripen from July through September. In North America, you will find the hardy red raspberry growing from Newfoundland to British Columbia, down through the Midwest, and south to North Carolina. It is especially prolific in California, where it boldly fruits into November, daunted only by the first frost. In the eastern states, 'Latham', an early-to-midseason mainstay, is the grandfather of cultivated red raspberries. In the West, you are more likely to encounter the dependable, sturdy midseason 'Willamette' and the two-crop 'Heritage', although the latter also grows well in the East. The red raspberry is equally hardy in northern Europe, because it thrives in cool climates. Red raspberry varieties often seen in the British kitchen garden are 'Malling Jewel', 'Glen Clova', and 'Malling Admiral'—easily harvested and resistant to blight.

Most cultivated red raspberries are the result of crosses between the Eurasian species *R. idaeus* and the American native *R. strigosus* and are so finely melded that they are usually grouped under the European nomenclature. Even botanists are perplexed by the red raspberry's comings and goings. Cultivated European raspberries escaped from New England gardens centuries ago; they mingled with native red raspberries and produced their own transatlantic lineage.

If you want to be savvy about American raspberries, do learn the distinction between the red raspberry and the purple-flowering raspberry, *R. odoratus*. The latter is a charming species that flourishes in country hedgerows in North America from Nova Scotia and Ontario south to Georgia and Tennessee; it does not have any hooked prickles, although its flower stalks are swathed in reddish bristly hairs. Its leaves are maple-shaped. Although they *are* edible, the fruits of the purple-flowering raspberry do not have the very sweet taste of other raspberries. This is no reason

to look askance at this plant, however, because it is delightful on its own terms.

The exquisite purple-flowering raspberry grows in a relaxed, straggling form and can climb as high as five feet. The fragrant flowers are a stunning lavender shade and could easily be mistaken for the wild pasture rose; they blossom continuously from June to August.

It's very pleasant to happen upon the flowering raspberry in full glory on a summer day, growing in a rocky forest or perhaps clinging to the side of a ravine. The flat, dome-shaped fruits look like they are covered with faded red velvet, so much so that you'll have to touch them to make sure they are real. Horticulturists who dismiss this berry as "worthless" are shockingly at fault, for they are not recognizing the unusual beauty of the plant. If you grow a purple-flowered raspberry plant in a semi-shady, moist location, it will produce ornamentally enticing fruits from July all the way through September.

A North American cousin of the flowering raspberry is the black raspberry, one of nature's more pleasant experiments. A curious anomaly, this lustrous fruit seems to crave anonymity. Many uninitiated berry pickers take it to be a blackberry; when unripe, it can also masquerade as a red or even yellow raspberry.

Nevertheless, there are many people who instantly recognize this ebony-berried plant and have bestowed a few pet names on it. It is sometimes evocatively termed *blackcap*, especially by mountain folk of Appalachia, because of the neat caplike shape of the berries. Appalachians put the berries to use in two ways: for light jellies and jams, they gather the berries just before they are ripe and are in a white or pinkish stage; for deeper-toned preserves and for wines and liqueurs, they harvest blackcaps when they are very ripe. Appalachians also dry the vitamin C–rich leaves of the blackcap and use them as tea. Steep the raspberry leaves for fifteen minutes or so, then strain them and add milk and sugar to taste to create a delicious and

The beautiful lavender blooms of the purple-flowering raspberry, opposite page, will decorate your property and are right at home in a rustic cottage-garden setting.

If you don't have the time or space to grow raspberries at home, the next best places to get them are the pick-your-own farm and the farmer's market, above. Remember to handle these fragile fruits carefully and keep the berries in a cool place.

warming concoction fit for a king or queen.

Another sobriquet of the black raspberry is *thimbleberry*: when picked, it separates easily from its receptacle, leaving a thimble-shaped hollow within. This conical berry's rich acid flavor made it a favorite of many Pacific Northwest tribes of American Indians; some people rate it the finest cooking raspberry available.

The black raspberry has strongly whitened canes, which take on a ghostly look in the berry patch, especially on misty mornings. Both the canes and the leafstalks are equipped with prickles that fiercely guard the fruit growing in neat rosettes. It grows from New Brunswick to Minnesota, but spreads down south into Georgia and west to California. The blackcap does not like to winter in the North, not being as cold-tolerant as red and yellow raspberries. A sun worshipper, it is able to withstand intense heat and is harvested prolifically in California well into October. Hybrids of black and red raspberries produce, not surprisingly, purple raspberries.

The rare golden raspberry is, botanically speaking, really the same species as the red raspberry. A result of crosses between reds and native yellow Asian raspberries, it has merely lost its pigmentation. But something even more wonderful happened as well. The flavor is infinitely sweeter than that of the red raspberry, especially when there's a pink champagne blush on the berry, so golden raspberries are especially ambrosial as dessert berries. Some people contend that there's no difference in taste. Others insist that the golden raspberry is unpleasantly sweet, but this is probably because they're astonished to see a yellow raspberry on their plates. Favored golden raspberry varieties for northern climates are 'Fallgold' and 'Amber'. Other relatives of the raspberry are discussed in the section on *Rubus varieties* (page 138).

Curiously, Isabella Beeton, in her *Book of Household Management* (1861), refers to two types of raspberries: red and white. The latter was the English white-fruited raspberry which was once grown in many gardens. Its place has been taken today by the golden raspberries. Mrs. Beeton felt that the raspberry is "exceedingly wholesome, and invaluable to people of a nervous or bilious temperament." The source of Mrs. Beeton's information is unknown—perhaps she observed pleasant personality changes in high-strung guests after eating her raspberry tarts—but you must agree that indulging in fresh raspberries has a particularly calming effect.

Two-crop golden raspberries add their own fall color to this upstate New York garden, where the surrounding foliage has begun to change, right.

Harvested from the plant, the golden raspberry, opposite page, resembles the cloudberry in color and general shape. However, the sweet fragrance and taste of the golden raspberry bear no resemblance to the more tart flavor of its cold-climate _Rubus_ cousin.

Crunchy Raspberry Muffins

This recipe lightly balances the tastes of raspberry and cinnamon and completely avoids the cake-like consistency to which so many muffins fall prey.

(Makes 12 muffins)

Batter:
75g/3 oz plain flour
75g/3 oz wholemeal flour
100g/4 oz granulated sugar
50g/2 oz light brown sugar
2 teaspoons baking powder
2 teaspoons cinnamon
1 egg, lightly beaten
100g/4 oz butter, melted
120ml/4 fl oz milk
150g/5 oz raspberries (preferably fresh, but defrosted frozen berries will work)
1 teaspoon grated lemon zest

Topping:
50g/2 oz light brown sugar
25g/1 oz plain flour
65g/2½ oz chopped nuts
25g/1 oz oatmeal
20g/¾ oz butter, melted

Batter: Preheat the oven to 180°C/350°F/gas mark 4. Grease 12 cups of a patty or muffin tin or line with paper cases. Sift the flours, granulated sugar, brown sugar, baking powder and cinnamon together into a mixing bowl. Combine the egg, melted butter and milk; add to the flour mixture all at once. Stir until just combined.

Quickly add the raspberries and lemon zest. Fill each cup two-thirds full.

Topping: Make the topping by combining all the topping ingredients. Sprinkle evenly over the top of each muffin. Bake until nicely browned, for about 20 to 25 minutes.

Breakfast Boboli

Dreaming of raspberries is said to augur a love affair. If you should be so lucky to summon up raspberries in slumber, you can seal your fate upon awakening by indulging in a raspberry breakfast. This recipe is a truly delicious concoction that can be thrown together on the spur of the moment.

(Serves 8)

*1 boboli crust (30-cm/12-inch
    prebaked Italian pizza bread*)*
100g/4 oz cream cheese, softened
*150g/5 oz raspberries (preferably fresh,
    but defrosted frozen berries will
    work**)*
1 tablespoon Grand Marnier
fresh mint

Preheat the oven to 230°C/450°F/gas mark 8. Place the boboli crust on a baking sheet. Spread the cream cheese on the boboli crust and bake for 10 minutes. Sprinkle the raspberries with the Grand Marnier. Remove the boboli from the oven and arrange the raspberries on top. Gar-

nish with fresh mint. Slice it pizza style and serve warm.

*Boboli is difficult to find. If your local specialty food shop doesn't carry it, make your own dough from a packet of pizza mix, following the manufacturer's instructions. Partially bake it until very lightly brown. If you are preparing the dough in advance, freeze or refrigerate it, depending on how long it will be before you make the dish. You can also use it immediately after lightly browning it. Simply remove it from the oven and place the ingredients on top of it, following the same directions for the boboli. This option also allows you to mould the dough into different shapes, such as a heart for Valentine's Day.

**If you use frozen raspberries, combine the raspberries and Grand Marnier and sprinkle over the crust prior to baking.

Red Raspberry Tart Studded with Golden Raspberries

This beautiful tart incorporates two types of raspberries.

(Makes 1 23-cm/9-inch tart)

Shortcrust Pastry with Nuts:
75g/3 oz almonds or pecans
150g/5 oz plain flour
2 tablespoons sugar
150g/5 oz cold butter, cut into 1-cm/
½-inch pieces
1 teaspoon vanilla or almond essence
4 teaspoons cold water
¼ teaspoon salt

Soured Cream Layer:
2 eggs
100g/4 oz sugar
225ml/8 fl oz soured cream
1 tablespoon lemon juice
pinch salt

Raspberry Curd:
900g/2 lb red raspberries
65g/2½ oz sugar
lemon juice to taste
2 whole eggs
2 egg yolks
75g/3 oz unsalted butter

Garnish:
300g/10 oz whole golden raspberries

Shortcrust Pastry with Nuts: Preheat the oven to 180°C/350°F/gas mark 4. Chop the nuts finely or grind in a food processor. Combine the flour, nuts and sugar in the bowl of a mixer, using a dough hook. Cut the butter into the flour mixture until it achieves the consistency of cornflour.

In a separate bowl, mix the essence and the water together. Add the salt and stir to dissolve.

With the mixer on medium speed, add the liquid mixture to the dry ingredients and beat just until the pastry comes together. Remove from the mixer and use your hands to knead it a bit. Shape the dough into a disc shape, cover with cling film and refrigerate for at least 2 hours before rolling out.

Roll out the dough into circle that is 30 cm/12 inches in diameter and 5mm/¼ inch thick. Use a small, sharp knife to trim off the ragged edges. Transfer to a 23-cm/9-inch flan tin with a 2.5-cm/1-inch removable side, and trim the edges. Refrigerate for 30 minutes.

Bake the pastry shell with beans or weights for about 25 minutes. Remove the weights and bake just until the shell is dry but not browned, about 3 more minutes. Allow to cool. Leave the oven on.

Soured Cream Layer: Combine all the ingredients and beat together, either in a mixer or by hand, until just combined. Strain through a fine-mesh sieve. Pour into the prebaked, cooled tart shell. It should be about half full. Bake in the oven for approximately 15 minutes, just until set. Do not brown. Cool on a rack.

Raspberry Curd: Purée the raspberries and strain through a fine-mesh sieve to remove seeds. Measure out 350ml/12 fl oz of purée. Heat in a small saucepan with the sugar and lemon juice, stirring until the sugar dissolves.

Whisk the whole eggs and egg yolks together in a bowl. Whisk in one-third of the hot liquid to temper them, then transfer back to the saucepan with the rest of the liquid and cook over low heat until the mixture has thickened, being careful not to scramble the eggs. It should fall off the spoon in clumps. Remove from the heat, add the butter and whisk in to combine. Strain through a fine-mesh sieve.

While the curd is still warm but not hot, pour into the pastry shell. Use a spatula to smooth the top.

Decorate the tart with whole golden raspberries, set stem side down. Chill for about 30 minutes to set.

Barbecued Duck Breast with Raspberry Sauce

Raspberry duck is especially delicious when barbecued. You can use an ordinary barbecue and charcoal or wood as fuel for this recipe.

(Serves 4)

Duck Stock: (yields 250ml/8 fl oz)
1 duck carcass
½ medium onion, diced in large pieces
1 medium carrot, diced in large pieces
1 stalk celery, diced in large pieces
3 bay leaves
120ml/4 fl oz red wine

Raspberry Duck:
4 duck breasts
salt and pepper to taste
450g/1 lb fresh raspberries
1 medium tomato, chopped
1 tablespoon tomato purée
3 tablespoons raspberry vinegar
1 bay leaf
a few sprigs of fresh thyme
a few stems of fresh parsley
50ml/2 fl oz duck stock (see below)
100g/4 oz sugar
120ml/4 fl oz water
25ml/1 fl oz cassis

Duck Stock: Preheat the oven to 230°C/450°F/gas mark 8. Roast the duck carcass in the oven until golden brown (about 15 minutes). Make a bed of the vegetables in a large saucepan and place the duck carcass on top. Cover the vegetables with water, add bay leaves and red wine, and simmer until the liquid is reduced to about 250ml/8 fl oz. Remove ingredients, strain through a sieve into a bowl, cover and refrigerate.

After about 4 hours, skim off the fat that has risen to the top of the stock.

Raspberry Duck: Season the duck breasts with salt and pepper and barbecue for 8 to 10 minutes per side; remove from the barbecue and keep warm.

Place the raspberries, tomato, tomato purée, vinegar and herbs in a large saucepan. Simmer over low heat until the juice of the berries is released.

Pour the raspberry mixture into a sieve and press down with a rubber spatula to release all the juices back into the saucepan. Place the saucepan on the heat, add the duck stock and simmer.

Prepare a simple syrup by boiling the sugar and water for 5 to 6 minutes. Slowly add the syrup to the sauce until a slightly sweet flavour develops (not all of the syrup needs to be added). Add cassis and continue to simmer for 5 minutes. Pour through a wire sieve to remove all seeds.

Ladle the sauce onto a plate, place a sliced duck breast on top and garnish with fresh raspberries.

Creamy Raspberry Mousse

This gorgeous dish contrasts the brilliant colouring of red and golden raspberries as well as their subtly different tastes.

(Serves 8)

350ml/12 fl oz purée of fresh or frozen
    raspberries
2 tablespoons gelatine
3 tablespoons framboise or Grand
    Marnier liqueur
2 eggs, separated
6 tablespoons sugar
250ml/8 fl oz double cream
50ml/2 fl oz crème de cassis
300g/10 oz red raspberries
300g/10 oz golden raspberries
8 sprigs fresh mint

Strain the purée to remove seeds. Remove 50ml/2 fl oz for the sauce. Dissolve the gelatine in the liqueur. Warm in a small saucepan over low heat to dissolve. Stir into the berry purée.

Place the egg yolks in an electric mixer bowl. Add 2 tablespoons of sugar to the yolks and whisk over hot water until light in colour. Continue whisking off heat until thick and cool and then fold this into the berry purée.

Whip the cream, adding 2 tablespoons of sugar, and continue until stiff; fold into the purée. Whip the egg whites, adding the remaining sugar until stiff; fold mixture into the purée.

Refrigerate until firm. Add the crème de cassis to the reserved berry purée for the sauce. To assemble, divide the sauce among large plates, scoop mousse onto it and decorate with fresh red and golden raspberries and mint.

Variation: If you wish, you can place the mousse in almond tuile cups, as pictured. Simply use any standard recipe for 8 almond tuile cups.

As a sophisticated public slowly warms to the golden raspberry, it is becoming a far more common crop on berry farms. Some berry growers have developed their own golden raspberry preserves.

Raspberry-Apple Harvest Tart

This is a lovely dish to serve in autumn, when raspberries and apples are both in their prime season.

(Serves 6 to 8)

Butter pastry:
100g/4 oz unsalted butter, frozen
3 tablespoons iced water
100g/4 oz flour
50g/2 oz sugar

Filling:
6 to 8 large baking apples
150g/5 oz raspberries
50g/2 oz flour
100g/4 oz sugar
1 teaspoon cinnamon

Crumb topping:
100g/4 oz unsalted butter, frozen and
  cut into pieces
2 tablespoons white sugar
150g/5 oz light brown sugar
1 teaspoon cinnamon

Butter pastry: Preheat the oven to 200°C/400°F/gas mark 6. Cut up the butter into 2.5-cm/1-inch slices. Put the butter, iced water and all other pastry ingredients in a food processor; blend until the mixture forms a ball. Roll out and put into a 23-cm/9-inch flan tin with removable ring. Place a piece of baking or waxed paper and weights or beans in the pastry shell (to hold down the pastry). Cook the pastry for 8 minutes, then remove the waxed paper and weights or beans; cook for 8 more minutes. Remove from the oven and make a few small holes in the pastry with a knife to let out the air. Leave the oven on if you are making the tart straightaway.

Filling: Peel and core the apples and cut into bite-size pieces. Place the apples in a bowl with the raspberries, flour, sugar and cinnamon and mix them together with your hands. Fill the pastry shell.

Crumb Topping: Place the ingredients in a food processor and pulse until little pebbles and grains appear. Top the tart with this mixture and put on a baking sheet (so it won't drip onto the oven). Bake at 200°C/400°F/gas mark 6 for 45 minutes or until the apples' juice has thickened and begins to bubble.

Raspberry Pavé

This recipe will not fail to delight your dinner guests. The flaky crust and light-tasting cake make a delectable base for luscious fresh raspberries, and a glaze of redcurrant and raspberry jams is a smooth complement to the crunchy pistachios that cover the dessert.

(Serves 8)

Pastry:
225g/8 oz butter
225g/8 oz plain flour
25g/1 oz icing sugar
pinch of salt
175g/6 oz raspberry jam

Cake:
3 eggs
100g/4 oz sugar
1 tablespoon orange juice
100g/4 oz flour
65g/2½ oz butter, melted
   (5 tablespoons)
175g/6 oz redcurrant jelly
900g/2 lb whole fresh raspberries
75g/3 oz chopped pistachio nuts

Pastry: Preheat the oven to 180°C/350°F/gas mark 4. Mix the ingredients in a food processor until incorporated. Refrigerate for 1 hour, covered by waxed paper. Roll into a 43-x10-cm/17-x4-inch rectangle and refrigerate again for 30 minutes. Bake until golden and refrigerate for another hour. Use a pastry brush to lightly coat it with raspberry jam.

Cake: Turn the oven down to 160°C/325°F/gas mark 3. Lightly mix together the eggs, sugar and orange juice in a bowl using a fork. Warm the mixture in a double boiler until almost hot. Remove from the heat and beat with a whisk until cold. Fold in the flour and then the butter.

Butter and flour a 43-x28-cm/17-x11-inch shallow baking tin. Pour the batter into the baking tin and spread it out evenly using a spatula. Bake until golden and leave the oven on.

Trim the cake to the size of the pastry. Place it on top of the crust and lightly glaze it with raspberry jam using a pastry brush.

Lay fresh raspberries close together on top of the cake and glaze with warmed redcurrant jelly. Toast the pistachios on a baking sheet in the oven for about 5 minutes. Remove from the oven and cover the sides of the pavé with them.

Berry Paint

Painting with berries is a lovely way to create a work of art when you serve any dessert—particularly a tart or a cake—that lends itself to the fruity accents of strawberries, raspberries or even kiwifruit. It's a contemporary twist on the medieval practice of using berries as pigments for actual paintings. Use Berry Paint to create these stunning patchwork effects with chocolate glaze and crème fraîche. Experiment with various combinations of berries to arrive at the tastes and colours for your very own berry paintings. Paint with berries by pouring the mixture into a plastic bottle with a cap top with a small opening so that you can release the berry paint in precise amounts. (Don't use an icing bag for this procedure as the mixture is too thin to be piped through one.)

(Makes 50-120ml/2-4 fl oz)

750ml/1¼ pt water
500g/1 lb 2 oz sugar
300g/10 oz berries
juice of ½ lemon

Make a simple syrup by boiling the water and sugar for 2 minutes in a saucepan. Pour all but 120–175ml/4–6 fl oz of the syrup into a storage container and refrigerate, covered, for up to 4 days (this can be used for other berry paints or whatever). Boil the berries in the saucepan with the lemon juice and the simple syrup for another minute, stirring constantly. Purée in a food processor. Push the mixture through a sieve.* Let cool.

*When making berry paint from kiwifruit, do not strain seeds through a sieve as they look lovely in the purée.

Raspberry and Zinfandel Poached Pears

Berries and fresh fruit is a simple yet stunning combination that anyone can prepare. This recipe is wonderfully elegant.

(Serves 5 to 10)

> 450g/1 lb red raspberries
> 750ml/1¼ pt Zinfandel
> 250ml/8 fl oz water
> 2 strips lemon zest
> 2 strips orange zest
> 5 peppercorns
> pinch cinnamon
> 100g/4 oz sugar
> lemon juice to taste (optional)
> 5 slightly underripe pears, peeled,
>   cored and halved
> crème fraîche

Purée 350g/12 oz raspberries and strain through a fine-mesh sieve to remove seeds. Reserve 100g/4 oz to use as garnish. Combine all the other ingredients except the pears and crème fraîche in a large saucepan that will accommodate the pears. Bring to the boil and reduce the heat to a simmer, stirring constantly. Taste for proper balance of acids and sugar, adding a bit more sugar or lemon juice, if necessary.

Add pears. Use a plate or towel to press down and ensure the pears are completely submerged. Simmer gently until the pears are just tender, about 30 minutes. If they are not too soft, let them cool in the liquid, If they have softened, remove and allow to cool, cut side down, on a baking sheet covered with non-stick paper.

Cook the poaching liquid over medium heat until it is reduced by about half. It should have the consistency of syrup. Chill in the refrigerator.

Pour the syrup onto individual plates. Arrange one or more pear halves decoratively on each plate.* Garnish with fresh berries and a dollop of slightly sweetened crème fraîche.

*As a variation, you can slice the pears and fan out the slices on a pan, cover them with crème fraîche, sprinkle them with sugar and grill them for 1 minute. Remove from the grill and place them in pools of poaching syrup on the plates and garnish with fresh berries.

ROWANBERRY
Sorbus aucuparia

Two hundred years ago in a northern European cottage kitchen, you would likely find a batch of rowanberry preserves simmering on the hearth. If it were near either the first day of May or of November, when witches and evil-minded fairies would be most likely to be prowling, there might also be a bough of the rowan tree hanging over the front doorway as a talisman to protect against evil spirits.

The rowan, or European mountain ash, as it is sometimes known, was once considered one of the most magical plants in the wood, and it was planted liberally in northern countries. (Don't be misled by its common name, as it is not even related to the common ash (*Fraxinus excelsior*), but is actually in the same family as hawthorns and apples.) This striking tree, a native of Europe and western Asia, can grow to fifty feet or more and yields sour-tasting orange or scarlet berries in late summer.

The name of the rowan in different languages reveals the deep esteem in which it is held by northern peoples. The Anglo-Saxon word for rowan was *wice*, close to the word for witch (*wicca*). The rowan was thought to protect against cantrips, or witches' spells. As the old British couplet states, "Roan-tree and red thread/Haud the witches a' in dread." The rowan is known by Swedes as *rönn*, which is most likely derived from the Scandinavian word *runa*, a rune or charm. And Ravdna, the ancient Finnish Lapps' thunder goddess, whose name is thought to be a cross between the Icelandic and Swedish words for rowan, favored the rowan so much that she cultivated groves of the beloved tree.

Finnish folk verse also pays frequent tribute to the rowan's berries and branches. An old Scandinavian-Russian belief is that if you are forced to spend the night in the woods, sleeping under a rowan tree or on a pillow of rowan boughs is the best protection against evil beings. Scandinavian folklore also endows rowan branches with mystical powers, and they were often used in combination with other woods as divining rods at sundown on Midsummer Eve to find precious treasure.

Once you have learned the formidable lore of the rowan you might think people would be too awed to harvest its berries. But its glossy brilliant red fruits grow so abundantly in northern regions of Europe that culinary usage was inevitable.

Rowan Jelly

The following recipe for rowan jelly is from Somerset. It is recommended as an accompaniment for lamb or venison.

(Yields 3 450g/1-lb jars)

800g/2 lbs rowanberries
2 or 3 peeled apples
200g/8 oz of sugar for each 240ml/
    8 fl oz of juice
lemon or lime zest

Place washed berries and peeled apples in a saucepan, crack the seeds, cover with water and simmer for a while, until the goodness has been extracted from the fruit and it is pulpy. This takes 20–40 minutes; the fruit must become quite soft. Put the mash into a jelly bag and let it drip into a bowl overnight. Don't prod or help the liquid in any way or it will get cloudy.

Measure the juice and add the sugar accordingly. Tie some lemon or lime zest in cheesecloth and add it to the saucepan (so that you can remove it easily). Dissolve the sugar slowly. Then boil until the jelly sets when tested. (Dip a cold spoon into the mixture. If the mixture runs off the spoon in a single sheet, as opposed to two drips, the jelly has set.) Pour into sterilised jars. Cover with melted paraffin, let cool, and seal.

The rowan, or European mountain ash, should be planted for both its edible and ornamental attributes. The rowanberry needs to be cooked before consumption; when raw, the taste is disagreeable.

RUBUS VARIETIES

BLACK SATIN

KING'S ACRE BERRY

OLALLIE

*S*panking new compared to such ancient berry cohorts as the strawberry and the raspberry, *Rubus* varieties are the results of plant breeders' attempts to improve on nature. The sheer enormity of the *Rubus* genus makes it possible for varieties to take on an endless array of features—characteristics that nature did not provide in the traditional form of the plant, such as the absence of thorns, larger berries, increased hardiness, or an early fruiting season. Differences in taste are often also significant, inspiring those with finely attuned palates to hold berry tastings of the same complexity reserved for wine, chocolate, and other foods that inspire obsession.

For the home berry gardener, the blackberry variety 'Black Satin' is an excellent choice. 'Black Satin' is a very young berry, yet its conduct is impeccable. It is a heavy yielder, thornless, produces extremely large berries on semierect canes, and has a long bearing season. The plants are very

vigorous and winter-hardy, with excellent disease resistance—which is the entire point of developing new varieties. This high-tech marvel was developed in America in 1974 and derives from very carefully chosen parents: ('US 1482' × 'Darrow') × 'Thornfree', the last contributing its tangy,

The 'King's Acre Berry', at left, is a distinctive turn-of-the-century hybrid that should be served at table with a regal flair, perhaps accompanied by a decorative berry spoon, a favorite utensil of the best Victorian households.

American hybrid 'Black Satin', right, is a late-model blackberry variety that is easy to harvest as it has no thorns. Both berries are pictured at the Royal Horticultural Society's Garden at Wisley.

usually quite large fruits and pleasant thornless disposition. The big, shiny black berries ripen on sturdy canes about two weeks before traditional blackberries and have a long bearing season. They have a very sharp flavor and a particularly enchanting aroma.

Another *Rubus* variety that gardeners favor is the 'King's Acre Berry,' which is on display at the Royal Horticultural Society's Garden at Wisley, England. The society honored this fruit with its Award of Merit in 1911, soon after Mr. J. Ward of Shobdon, Herefordshire, somehow produced the berry, of *Rubus* parents that are still a mystery. The fruits that resulted from Mr. Ward's successful, albeit off-the-record, experiment are quite sweet and make a sizeable, gratifying mouthful. The cane growth is moderate, and the plants are thorny, however, not thorny to the point of being rudely aggressive.

A *Rubus* variety with a bigger following is the famed 'Olallie' berry, the California kingpin of blackberries—which, incidentally, was developed at Oregon State University. The large contingent of 'Olallie' devotees doesn't even mind carefully reaching through this plant's painful thorns to get to the oversized, sturdy berries borne so prolifically on the canes. The 'Olallie' thrives in southern California, particularly in San Luis Obispo County, but has a very short growing season—from June through July. The berries tend to rot quickly, so many 'Olallie' aficionados freeze them immediately to preserve the nippy yet sweet taste for later in the year.

Some berry farms grow 'Olallie' berries in such abundance, left, that there are even berries left over for making juicy sweet pies, right. You might consider wearing gloves when picking 'Olallie' berries — the thorns are quite sharp.

Olallie Berry Pie

If you are fortunate enough to obtain the elusive 'Olallie', you are well advised to use the berry in this pie recipe. For optimal results, you must select only ripe, juicy berries picked at the height of the season. Not only are they more full of flavour, but they also require less sugar, which will give your pie a delicious tang. Use your favourite pastry recipe. Blackberries can also be used for this recipe.

(Serves 6 to 8)

Pastry for 2 23-cm/9-inch pie crusts

Filling:
300g/10 oz sugar
25g/1 oz cornflour
550g/1¼ lb 'Olallie' berries

Line a 23-cm/9-inch dish with one of the pie crusts. Preheat the oven to 220°C/ 425°F/gas mark 7. Mix the dry ingredients into the berries, and then place them into a pastry-lined pie dish. Cover with a top crust. If you wish, you can brush some milk on the top crust.

Bake for 45 minutes to an hour, until the pastry is nicely browned and juice begins to bubble through the crust. Allow to cool and serve *à la mode*.

STRAWBERRY

CULTIVATED STRAWBERRY
Fragaria ananassa

ALPINE STRAWBERRY
Fragaria vesca

The strawberry, like the fedora and the suntan, has slipped in and out of fashion over the years at a dizzying pace. Sweet as it is, the strawberry has not always been regarded in the best light. In fact, there is something of a curious strawberry dichotomy throughout history. On the one hand, it is a spiritual, noble fruit; on the other, like the forbidden apple, a tempting, sensual delight. The strawberry has been portrayed as the symbol of deities as disparate as the Virgin Mary and Frigga, the Teutonic love goddess.

Fifteenth-century religious miniature paintings often depicted garden scenes to represent abstract spiritual values. Different plants signified specific virtues. The strawberry represented the fruits of righteousness and its leaves the Trinity. In the early sixteenth century we again find the strawberry in art, but cast in a completely different role.

In Hieronymus Bosch's triptych *Garden of Earthly Delights*, strawberries are synonymous with carnal pleasure: evil creatures gorge themselves on the carmine fruits; some members of this wicked horde even have strawberries as body parts.

The household records of the English King Henry VIII further the association of strawberries with debauchery, indicating that he had a fondness for the exorbitantly priced berries, which were probably *Fragaria vesca*. One imagines the bloated King Henry reclining in his elaborate padded-sleeved costumes, draped in a fur-lined robe, devouring bowls full of the dainty berries. As strawberries were an old remedy for gout, from which he suffered, perhaps this was the motivation for his berry consumption. But more likely than not, he was enthralled, as we all are, by the berry's splendid taste.

Strawberries, despite Henry's example, were not especially valued as a curative. In fact, in the sixteenth century, doctors advised their patients not to ingest too many fruits or vegetables. It was believed that they were unhealthy if eaten in abundance.

It's not surprising then that John Gerard, a sixteenth-century surgeon and herbalist who cultivated more than two thousand plants in his London garden, remarked little on the strawberry's excellent taste in his famous *The Herball*. Among the "vertues" of strawberries that he *did* mention, however, were that the leaves soothed wounds, strengthened the gums, and "fastneth the teeth." He also recommended distilling strawberry water and drinking it with white wine. This concoction, he maintained "is good against the passion of the heart, reviving the spirits, and making the heart merry." The only recommendation Gerard could make for fresh strawberries is that they "quench thirst" and can also take away the "redness and heate of the face" if used on a regular basis.

Despite its medicinal reputation, the strawberry was still eaten surreptitiously for pleasure by an enlightened few. Strawberries were one of the first packaged foods, sold in the sixteenth century in pottles, cone-shaped straw baskets. Because they had to be harvested, packaged, and brought to market as speedily as possible, they were rather expensive, which contributed to their reputation of being a snobbish food consumed by the upper classes, much in the same way caviar is regarded today.

Even as late as Victorian times, strawberries still commanded a high price and were viewed somewhat disdainfully by practical-minded folks. A manager of a settlement house in Hartford, Connecticut, noted with disapproval that a female factory worker living under his roof purchased two boxes of strawberries for almost a dollar—a full day's pay!—simply because she wanted to indulge herself.

The popular alpine strawberry variety 'Alexandria' is often used as an edging plant in the garden. Here it grows in a model backyard garden at the Royal Horticultural Society's Garden at Wisley.

143

Today the strawberry is a commonly grown fruit, and few would guess that it has such a checkered past. To make matters even more complicated, the strawberry's botanical history is as tumultuous as its social one. The contemporary cultivated strawberry is actually a very new addition to the *Fragaria* genus. It seems that in the early 1700s, while on a spying mission of unknown purpose in South America, a Frenchman named Amédée François Frézier became intrigued by the deep red, large-fruited strawberries of Chile. He smuggled the species, *F. chiloensis* (sometimes known as *beach strawberry*), to France, where it was introduced to the king's gardens in Paris.

Once on the Continent, it was inevitable that this South American species should meet with its North American counterpart, *F. virginiana*, which had been previously sent to England by the Virginian colonists and had met with great acclaim for its large size and prolific fruiting. Another Frenchman, Antoine Nicolas Duchesne, took a special interest in these New World species. When he was only seventeen years old, he presented King Louis XV with a pot of extremely beautiful and large *F. chiloensis* strawberries pollinated by *F. moschata*. The young man was commissioned by the king to raise the plants of Versailles and to also collect all the strawberry species known in the world. Although the exact origin of the parent plant of the modern cultivated strawberry is debatable (some writers report that Duchesne was this berry's creator), Duchesne was the first to identify the crossed parentage of *F. virginiana* with *chiloensis*. He catalogued several *F. ananassa* varieties in the 1771 supplement to his *L'Histoire Naturelle des Fraisiers*, first published in 1776, when he was nineteen years old. Today's familiar garden strawberry, *F. ananassa* is a direct descendant of these first fateful crosses of the eighteenth century.

After this perfect berry had been created, strawberry breeding became commonplace. The Victorians frenetically

Where there are strawberries, there is mulch. In addition to protecting strawberry plants in winter, it prevents the spread of weeds and helps the soil retain moisture during the fruiting season.

attempted to breed berries whose tastes were variations on the flavors of pineapples, apples, apricots, cherries, grapes, mulberries, and raspberries. Much attention was focused on the strawberry, and "Strawberry Wars" among American horticultural societies raged between 1842 and 1855. At the heart of their arguments was the question of whether there is a differentiation of sexes in strawberries (which, of course, there is).

If the idea of a genetically engineered berry disturbs you, it is time that you tried the wild strawberry. Unsullied by human intervention, wild strawberries—also known as *fraises des bois* and *alpine*, *wood*, and *heart berries*—are tiny, intensely sweet strawberries that have been with us all along, growing naturally in Europe (botanists differ on whether *F. vesca* is a North American native). Their heady aroma and flavor have lured berry pickers into woodlands for centuries. One of the first records of them in cultivation is under England's Edward I (reigned 1272–1307), who had the little strawberries transplanted from the wild to his castle garden.

Alpines will shower you with fruit from July until the first frosts. There is even a ghostly-looking white-fruiting wild strawberry, which has an unusual, almost tangy taste, but it is hard to come by. Two types of white alpines are 'Alpine Yellow' and 'Pineapple Crush'. Birds, usually strawberry pests, don't seem to find the whitish-yellow coloration very appetizing, perhaps because the berries don't look ripe. Easier to procure and quite satisfying to grow is the European red conical alpine 'Alexandria'. This variety doesn't form runners, so it makes a tidy border or ground cover.

Because they are such compact plants, alpine strawberries can easily be mixed with flowers to create a cottage garden effect. If you prefer a more formal style, use alpine strawberries as an edging plant in a garden to define a plot of vegetables. Many dedicated strawberry gardeners ensure profuse blooming and fruiting by adding a bee skep, or hive, to the center of a garden edged with alpine strawberries. Bees will pollinate the flowers, which in turn produce the fruit. You can also plant them in terraced beds or as borders along a sidewalk or driveway.

There are strawberries adapted to growing conditions in every part of the world—from the Scandinavian Lapland to the equator. Strawberries are summer's first fruits and actually begin decorating the garden in May, when the plants' delicate white flowers open, luring bumblebees to drink their nectar. By mid- to late June in most regions, strawberries begin ripening.

If you are puzzled by the vast variety of strawberries available, rest assured that there is a strawberry to suit your garden. Everbearing varieties produce toward the end of the first summer after planting. June bearers skip a summer following planting, while alpine varieties bloom from summer through autumn. Your local nursery carries varieties of strawberries that grow in your region; it is up to you to decide on such factors as color, size, and fruiting period.

Because frost damage to the strawberry plant can prevent it from fruiting, many picturesque and elaborate methods for caring for the plant have been devised over the centuries. In the seventeenth century, gardeners used straw "hats," shaped like bells, to prevent frost damage. In the Victorian era, glass cloches were used for the same purpose, and some Old World growers still adhere to this method. Contemporary growers usually protect the strawberry plants from frost in the winter by heaping straw mulch over them.

In the late Victorian era, strawberry houses consisting of layers of frames stacked on top of one another purportedly were developed. These shrinelike structures were meant to function somewhat like greenhouses to protect strawberry plants from cold, although period diagrams never explained exactly how they actually worked.

If you can appreciate the strawberry's scent as much as its taste, you should try growing it in the fragrance garden. Create

an herb garden and let the strawberry take its rightful place as an herb within it. An inventive gardener in upstate New York has created one such paradise. Beautiful and aromatic, this charming mix of a kitchen and cottage garden combines strawberries, flowers, and scented-leaf plants. Herbal plantings include lemon verbena (*Lippia citriodora*), bergamot (*Monarda* spp.), mints (*Mentha* spp.) and, winter and summer savories (*Satureja* spp.). She also has roses (*Rosa* spp.), snapdragons (*Antirrhinum* spp.), and pinks (*Dianthus* spp.) as well as a few night-blooming plants—flowering tobacco (*Nicotiana* spp.), evening stock (*Matthiola bicornis*), and dame's-rocket (*Hesperis matronalis*)—in her garden. The effect is positively exhilarating, nature's form of "aroma therapy."

Of course, the whole point of growing these wonderful berries is to eat them. Having a steady supply of fresh strawberries on hand will allow you to improvise with these versatile culinary jewels. Strawberries are enchanting additions to meals at any time of day and on any occasion. Their very presence in a meal instantly lends the repast a classical air. They make simple shortcake splendid, adorn tarts in a glistening glaze, and can be floated on warm Finnish cheese and cream—a traditional Scandinavian midsummer supper. Strawberries can also be dipped in chocolate, pureed as a dessert topping, sliced into a salad, strewn atop hotcakes, and mixed with liqueur to create a heady after-dinner potion. Or throw formality to the wind and indulge in the simple ambrosia of strawberries and cream.

When harvesting your strawberries, pick enough so that you can enjoy some now and preserve others for colder months. Memories of your summer strawberry picking will warm and lighten the most blustery days of winter as you gaze into the garden, where the berry plants lie dormant. If you find yourself particularly downtrodden by the weather, it's time to bring out the berries you preserved in the height of summer for enjoying before the winter hearth.

Growing only up to ten inches high, albino alpines are enchanting, precious additions to the berry garden or flower beds.

146

Savoury Strawberry Tartlets

This delightful appetizer takes advantage of the wonderful blend of sweet strawberry, tangy goat's cheese, and tasty Westphalian ham.

(Makes 12 tartlets)

> 75g/3 oz unsalted butter, cut into bits, kept cold
> 175g/6 oz flour
> pinch of salt
> 100g/4 oz Westphalian or other well-flavoured ham
> 150g/5 oz goat's cheese
> 4 to 5 large strawberries, quartered

Mix together the butter, flour and salt until all the flour is incorporated into the butter. Add a few drops of cold water until the pastry holds firm. Refrigerate for about 1 hour. Preheat the oven to 180°C/350°F/gas mark 4.

Roll out the pastry to about 3–5mm/⅛–¼ inch thickness and press into 4-cm/1½-inch round or square individual pie tins. Bake for about 25 minutes, or until golden brown. Allow to cool. Set the grill to high.

Tear or cut pieces of ham to fit inside the tartlets, with one or two layers of ham per tartlet. Place a pinch of goat's cheese in each tartlet and cook the tartlets under the grill for a minute, just until the cheese begins to melt. Place the strawberries into the softened cheese.

Five-Fruit Compote with a Mint and Tea Infusion

A classic European way to serve berries is in a compote, a dish featuring fruit cooked in a syrup. This variation on the traditional all-fruit compote is made by adding an invigorating blend of mint and tea.

(Serves 4 to 6)

Tea Infusion:
120ml/4 fl oz water
2 tea bags, preferably English Breakfast tea
3 large sprigs fresh mint

Fruits and Berries:
12 ripe large red strawberries
225g/8 oz black cherries (Bing, if possible)
1 ripe mango
150g/5 oz blueberries or bilberries
150g/5 oz raspberries

Syrup:
150g/5 oz sugar
2 vanilla pods, split in half, or use 1 teaspoon vanilla essence
775ml/26 fl oz dry white wine
120ml/4 fl oz port wine

Tea Infusion: Bring the water to the boil and add the tea bags and fresh mint sprigs. Stir until the mint is wilted. Remove from the heat and cover. Allow to stand 10 minutes or longer.

Fruits and Berries: Meanwhile, trim and discard the stem of each strawberry, leaving only the firm red flesh. Cut each strawberry in half lengthwise. Put the strawberries into a saucepan.

Pull off and discard the stem of each cherry. Cut each cherry in half and discard the stones. Add the cherries in with the strawberries.

Peel the mango. Cut the flesh into thin slices. Cut the slices into thin lengthwise strips. Add to these the fruits and berries.

Rinse the blueberries and raspberries and add them to the fruits and berries.

Syrup: Put the sugar in a saucepan and add the vanilla, white wine and port. Line the saucepan with a sieve and pour the tea mixture into the sieve. Squeeze each tea to release the liquid. Reserve the mint sprigs. Discard the bags.

Bring the mixture to the boil and pour it over the fruits and berries. Return the fruits to the pan and quickly bring to a simmer over high heat. Add the mint sprigs and remove from the heat. Place in the refrigerator and chill well.

Serve the fruits in shallow soup plates garnished with sprigs of fresh mint.

Strawberry Shortcake

Although it's a traditional dessert, strawberry shortcake need not be presented as such. This recipe breaks the usual mould and turns the shortcake into a lovely crescent moon.

(Makes 6 to 8 shortcakes)

Strawberry Mixture:
1.5kg/3 lb strawberries
50g/2 oz caster sugar
2 tablespoons triple sec

Shortcake:
300g/10 oz unbleached strong flour
2 tablespoons baking powder
2 tablespoons granulated sugar
175g/6 oz salted butter, cut into small pieces
120ml/4 fl oz soured cream
2 large eggs, lightly beaten
1 teaspoon vanilla essence
whipped cream
icing sugar

Strawberry Mixture: Take 350g/12 oz of the strawberries, the sugar and triple sec and strain through a sieve. Place the remaining fresh strawberries in the mixture and marinate for 1 hour.

Shortcake: Preheat the oven to 230°C/450°F/gas mark 8. Sift the flour with the baking powder and sugar. Cut the butter into this mixture. Make a well in the centre and add the soured cream, eggs and vanilla. Blend well with your fingers.

Turn the pastry out onto a floured board. Knead for about 1 minute and then pat out to 1-cm/½-inch thickness. Using a biscuit cutter about 7.5-cm/3 inches in diameter, cut the pastry into half-moon shapes. Scrape up the remaining dough and again roll out to 1-cm/½-inch thickness until you've used up all the dough. You should end up with 12 to 16 half moons. Place these on one (or more, if the pastry doesn't

fit on one) clean and dry baking sheet and bake for 15 minutes.

Cut the shortcakes in half. Spread the strawberry mixture on the bottom half and top with whipped cream. Add a top half of shortcake, and top with the mixture and fresh strawberries and whipped cream. Dust with icing sugar.

Alpine strawberries are so intensely flavorful that just a few added fresh to a meal can completely change its character, as demonstrated in this dish by chef Michael McClernon of the restaurant Paloma. A study in elegant simplicity, this quail entree is garnished with alpines. The berries also impart flavor to a glass of champagne.

Chapter Four

BERRY GIFTS:
PRESERVING, DRYING, AND STORING EDIBLE BERRIES

iving berries as gifts is one of the most thoughtful gestures you can make. Not only will lucky recipients be appreciative of the care that went into creating their gifts, but they will also be delighted by the delicious contents. Best, these gifts are appropriate for any occasion and are inexpensive yet attractive.

Making preserves at home is a satisfying pastime, especially when you can share the results with friends. Preserved berries packed in attractive glass jars make stunning gifts. Save jars that are too lovely to discard, heap them with berry preserves, and festoon them with ribbons for special friends; these are especially nice presents for the holiday season.

Such standard tomes as *Jane Grigson's Fruit Book* are recommended reading for the home berry preserver. But making fresh berries into luscious jams and jellies is not all that difficult. Generally speaking, cook jam or jelly to a desired thickness using your favorite recipe and use a wide-mouthed funnel to ladle or pour preserves into jars. Fill jars to within one-quarter inch of the top. Then seal with a sterilized lid. Use paraffin or flat lids to seal jars. Afterward, use a jar lifter to transfer the jars to a rack. Let them cool.

Aside from turning them into jams and jellies, you can make fresh berries into other marvelous concoctions. Add Madeira wine to a jar filled with blackberries or raspberries and let the mixture stand for at least two weeks. This luscious sauce can then be poured over cakes for a truly memorable dessert. It can also work as an accompaniment for game or meat.

Of course, if you live in an area with a sunny, dry climate, there's also the option of sun-drying berries to preserve them, as the North American Indians did centuries ago. In *Putting Food By*, Janet Greene recommends putting the berries in cheesecloth and dipping them in rapidly boiling water for fifteen to thirty seconds and then plunging them into cold water before you

Berries — and their leaves and flowers — can be enjoyed in potpourris. Pictured here is strawberry potpourri in the large basket and bayberry spice and woodsy rose potpourris in the jars. Placed in miniature baskets or glass jars, these potpourris are thoughtful, long-lived gifts.

begin the process of drying the berries.

Berries can also be sun-cooked, given sunny weather, as they are at the Green Briar Jam Kitchen in East Sandwich, Massachusetts. Sun cooking is a turn-of-the-century method of heating berries and sugar on a stove over low heat until sugar melts and reaches a full boil, pouring the mixture into pans, and letting it sit for two or three days under a window in summer. The result is a thick sauce brimming with delicious, sweet whole berries.

To sun-dry fresh berries, arrange them one layer deep on screens or racks, which allow for air circulation. Place the berry-filled racks outdoors in the sun and cover the berries with cheesecloth. Leave the racks outside during the day and bring them in at night. Do this for a week of hot, sunny days. If you live in a climate where a week of sunny days is a rarity, you *can* dry berries in your oven or in an indoor dryer. Consult books on preserving food (see the Bibliography), for specific sun-drying procedures to follow.

Putting Food By also recommends pasteurizing fruits after drying them by putting them on trays up to one inch deep in a 175-degree oven for ten to fifteen minutes. They can then be added to sauces, pancakes, pies, muffins, and ice cream in winter when fresh berries are unavailable. Firm berries such as the bush berries—currants, gooseberries, and blueberries—and cranberries and huckleberries dry best. When the berries are dried, store them in airtight containers.

Home brewing of berry wines is another possibility for the adventurous gift giver and entertainer. Gooseberry wine is a surprisingly sparkling beverage; it goes particularly well with fish. Currants make excellent dessert wines. A Pacific northwesterner who brews this beverage reports that he can't make enough to satisfy his clientele. The blackberry has a mellow flavor that is very pleasant as a table wine.

Vinegars are another method of preserving berries. The fruits (raspberries and blueberries work particularly well) are steeped in vinegar for several weeks in jars. The berry vinegars can be used in myriad ways: sprinkled over sliced fruit, basted on chicken, mixed with jam as an accompaniment to chicken or duck, or made into vinaigrettes. Berry syrups and butters are other food gifts that you can create. Raspberry butter spread on pancakes is a rare and delectable treat.

You may wish to preserve your freshly picked berries for future use simply as they are. One method of freezing berries is to spread them one layer thick on a cookie sheet and place it in your freezer. After the berries are frozen, pack them in containers or freezer bags. When you're ready to use them, thaw the berries and prepare them as you would freshly picked types. If you want to freeze berries in sugar, a general rule of thumb is to add ¾ cup for each quart of berries and mix the sugar through them. Place them in a container and seal. When freezing strawberries, do not remove the caps. You will probably want to use your frozen berries within six months, but most will keep for some time, some for up to a year.

After you have picked berries, you should place them in your refrigerator as soon as possible, uncovered and unwashed on a tray or plate. Eat them fresh or preserve them within two to three days for best results. Wash and drain them gently just before using. Never allow berries to stand in water and always handle them carefully to avoid bruising them.

There are many uses for other parts of the berry plant. Strawberry, raspberry, and blackberry leaf teas are popular by-products. It's best to obtain such concoctions commercially (see "Sources"), as the leaves can become poisonous if contaminated by moulds during the drying process.

Few foods are as versatile as the berry. With so many uses and so many ways to preserve them, it's possible to enjoy berries year-round. Don't forget that one of the most appreciated gifts of all is a simple basket brimming with an assortment of these freshly picked fruits.

Berries can be preserved in countless ways, but vinegars are among the most versatile and aromatic methods. Pictured are red and golden raspberries soaking in vinegar. A wonderful gift for a garlic lover is giant cloves steeped in red raspberry vinegar.

Strawberry-Peach Jam

This gorgeous jam, made from fruits plucked at the height of summer, has a beautiful red colour. Its taste is pleasantly evocative of warmer, more carefree days on wintry mornings.

(Yields approximately 9 175-g/6-oz jars)

550g/1¼ lb hulled strawberries
900g/2 lb peaches, peeled, pitted and
  cut into bite-size pieces
1.25kg/2½ lb sugar

Prepare the fruit. Place in a very large aluminum pan and stir in the sugar. Bring the mixture to the boil; this will take 5 to 10 minutes. Reduce heat to maintain the boil and cook until the mixture has thickened and setting point is reached, about 20 to 25 minutes. You'll be able to tell the jam is ready when the mixture drops off a cold spoon in a single sheet rather than in individual drops. Pour into sterilized jars. While the jam is still hot, remove air bubbles that rise to the tops of the jars by going around the edges and lifting them with a knife. Wipe jar covers with a cheesecloth dipped in water to clean them and screw on covers. Seal with paraffin. The jam has a very long shelf life, but try to use it within 6 months for the freshest taste.

Sun-Cooked Strawberries

Berries cooked in the sun have a mellowed sweet taste. Use these sun-cooked strawberries as a dessert topping—spooned over ice cream or poured over plain cake. You can cook your strawberries under a frame so they get sun from dawn to dusk, so there is no need to remove the berries at night, but if you sun-cook your berries in your garden, remember to take them in at night.

(Yields approximately 4 225-g/8-oz jars)

900g/2 lb strawberries
1.5kg/3 lb sugar

Carefully wash and hull the strawberries. Place them in a very large saucepan with the sugar and cook them over low heat until the sugar is melted, about 10 minutes. (Don't shake the pan as you want the berries to remain whole and unbroken. Do not even stir them.) Raise the heat and cook until the mixture comes to a full boil, about 5 to 10 minutes. Pour into a shallow bowl or pan—a 23-x33-cm/9-x13-inch Pyrex pan or 1.7-l/3-pt enamel pan. Remove any foam by skimming it from the top using a slotted spoon. Cover with a piece of glass and place it in full sun for 3 to 5 days. You'll know the mixture is done when it reaches the consistency of a thick syrup and has a sheen when it is tilted slightly.

Fill sterilized jars to the top with the syrup and let them sit overnight, unrefrigerated and covered loosely by paper towel. In the morning, remove air bubbles that have risen to the top of the syrup using a knife. Simply move the knife around the edges of the jars to lift out the bubbles. Seal lids with paraffin to preserve. The jars have a shelf life of about 6 months.

When the clock chimes four, it's time for tea — and berries, too. Spread delicious berry jam on tea cakes or biscuits, or serve berries fresh with cream. This melange of preserves, left, includes strawberries, raspberries, and beach plums.

If you want to give a gift that will be the talk of the dinner party, find a strawberry patch, above, pick a brimming basketful, and present it to your host or hostess with a bright bow attached.

Chapter Five

ORNAMENTAL BERRIES

INTRODUCTION

THE ORNAMENTAL BERRY GARDEN

When the lush green and bright floral colors of summer fade, there's no reason to despair. Another, more textured landscape unfolds, with a palette of colors ranging from alabaster to umber. As the summer garden quietly drops its petals and leaves, the ornamental berry garden emerges in all of its glorious finery to begin a two-season reign.

Plants bear berries after they flower. Ornamental berries are those we treasure principally or purely for aesthetics. While they may be eaten by wildlife, they may either be poisonous or insipid-tasting to humans. Some are indeed edible, such as rosehips and some species of barberry and sumac, but their principal usage among the majority of people is visual.

Most ornamental berries ripen in autumn after a late flowering period, and many remain on the plant into winter. Because they fruit after strawberries, blackberries, raspberries, and other appealing summer fare consumed by humans and wildlife, they do not have to compete with the edible berries for animals to spread their seeds. Nature arranged it that way to

ensure their survival. Some ornamental berry plants fruit concurrently with the edible berries, but their fruits may persist into autumn—and sometimes even winter—because they are not harvested by people and animals are slow to eat them. Something inherent in the plant determines when the seeds are dispersed. For example, *Symplocos* berries appeal strongly to birds and can easily be plucked from the branch. The 'Sparkleberry' cultivar of *Ilex verticillata*, by contrast, holds onto its fruits tenaciously for a long time; birds simply can't remove them easily from the shrub. With these many factors at play, you must carefully consider which plants to grow in your ornamental berry garden.

Designing an ornamental berry garden is as much an exercise in logic as it is in aesthetics. Consult any standard, up-to-date horticultural reference book, such as *Hortus* or Michael Dirr's *Manual of Woody Landscape Plants* for America; Hilliers' *Manual of Trees and Shrubs* for Great Britain, to find out what species will grow in your area. After you have narrowed down the choices, consider the palette of berry colors available and their fruiting periods.

The scarlet and black berries of *Berberis*, flowering dogwood, firethorn, the rugosa rose, and viburnums are gorgeous in the autumn garden. Orange-yellow berries of bittersweet and lilac beautyberries complete the spectrum. Also consider the cotoneasters. Depending on the species, you have your choice of orange or red berries of many sizes. Among the hawthorns, you can have a continuous red berry and black berry display from early autumn through spring. Consider training the rock spray cotoneaster (*C. horizontalis*) along a wall or rocky area of your property. Of course, there are also the higher-growing types, such as *C. racemiflorus*.

If you want your ornamental berry garden to have a natural, unstudied appearance, simply look to the woods for suggestions. Observe what combinations of plants grow naturally in a symbiotic relationship in your area and plan your garden accordingly. A basic, reliable plan is to create a border, with low-, medium-, and high-growing plants situated so that the view of one plant isn't obstructed by another plant. A tree or two can stand on a lawn in front of the border, perhaps under-

At this lovely private garden in South Petherton, England, _Skimmia japonica's_ brilliant red berries are a beautiful landscape accent and draw the garden visitor down a pathway.

Barberry makes its home on several continents. Here, left, it ornaments the grounds of an old boathouse on the Hudson River.

Ornamental berries can perform many functions. They can form hedges, act as specimen plantings, or grow as ground covers, as pictured above.

planted with a ground cover. If you enjoy evergreen color, look for this feature in the berry-bearing shrubs you're considering planting. Don't think that simply because some members of a genus are evergreen, all members are. For example, we are all familiar with the evergreen hollies, but there are also deciduous types that have berries. Some people prefer deciduous berried plants because the berries stand out so much more starkly on the plants.

Should you wish to introduce garden structures, consider a pergola planted with a twining vine such as bittersweet or Virginia creeper. If you are a bird lover, the ornamental berry garden will provide you with many hours of enjoyment, as birds dart and swoop at bright red honeysuckle berries or copper-colored sorbus berries.

Also when planning your garden, visit local botanical gardens in winter to see what they offer. You may be surprised, for example, to see the lavender berries of beautyberry, *Callicarpa japonica*, glowing brightly in the middle of December as the temperature dips to below freezing. Indeed, such delicate-looking berries as the snowberries (*Symphoricarpos*) and transparent-berried mistletoe honeysuckle (*Lonicera quinquelocularis*) persist into the long winter months.

Ornamental berries have been utilized throughout history for more than their landscape beauty alone. They have inspired elaborate customs, such as decking halls with holly and mistletoe. North American Indians made use of their leaves, bark, roots, and berries for medicinal purposes. In northern countries, barberries, the berries of Saint-John's-wort, and honeysuckle berries are ascribed magical powers. Many berries have been used by European and American households as clothing dyes and inks. You will, no doubt, find your own uses for them—in flower arrangements, wreaths, garlands, centerpieces, and other delightful decorations that can brighten the home even during the coldest days of winter.

If you have a passion for purple, beautyberry's resplendent winter display of color is sure to please. Children are particularly captivated by the lovely berries, which resemble hundreds of beads strung together. It is pictured in December at the Brooklyn Botanic Garden.

Berries are as diverse as flowers in color range. Lavender beautyberries, above, are stunning in the landscape and are also enjoyed by birds in the depth of winter.

Ornamental berry-bearing plants can be used in any garden style. In an informal British garden, berries spill out over a pathway, right.

Do not be put off by the baneberry's suggestive name. If you prefer, call baneberry by its other common names, chinaberry and doll's eyes. Although poison to the palate, it brightens the garden with its white, chickpealike berries, above, which are also good for flower arrangements.

BARBERRY

COMMON BARBERRY
Berberis vulgaris

JAPANESE BARBERRY
Berberis thunbergii

*V*enerable barberry, whether growing untamed or pruned meticulously in the knot garden, has weathered the centuries gracefully. Old as the Rosetta stone, yet blooming anew each year, barberry is the craggy garden favorite of the Scottish witch, the tonic of the American Indian, the plague preventative of ancient Egypt. Most people know barberry by its autumn fruits, which differ in color among species, from red to yellow to black, red being most common.

Barberry is actually considered an herb because of its purported medicinal properties. In the medieval period, barberry-derived medicines were used as astringents and cure-alls. John Gerard noted in *The Herball* that the barberry works well "against hot burnings and cholerick agues," finding it "also profitable for hot Laskes, and for the bloudy flix." Even if it's difficult to discern the nature of the ailments barberry was reputed to cure, Gerard's claims certainly sound impressive.

There is evidence to back up the barberry's alleged medicinal properties. A substance found in the roots and leaves, berberine, is thought to produce a whole range of physical benefits, working as both an anesthetic and a stimulant. The Nepal barberry, *Berberis aristata*, has inspired some interesting uses. The roots and wood are used in extract form as a tonic, astringent, and purgative. In India, where a common name for barberry translates as "ophthalmic berry," a concoction made from the root bark and applied over the eyes to relieve pain is a widespread household remedy; of course, the two parts opium that comprise the mixture probably play a role in numbing any discomfort. Studies are still inconclusive, however, so it is most definitely unwise to experiment.

Some people prize barberry for its beauty alone. The nineteenth-century garden writer William Robinson wrote in his classic *The Wild Garden*, "The prettiest brake of shrub I ever saw was an immense group of the common Barberry, laden with berries weeping down in glowing color." Robinson was referring to *B. vulgaris*, the common barberry of British lore and culinary usage. But other barberries have much to recommend them.

The Japanese barberry, *B. thunbergii*, a very familiar sight in North America, is a handsome compact shrub with lustrous red berries. When winter arrives, you will take delight in admiring its glowing red berries in the landscape, weaving color into a landscape barren of color but full of texture. There are barberries adapted to Europe, Asia, Africa, and North and South America, so it is likely that you will have a number of species to choose from for your particular climate. The only drawback to growing the plant is that some forms are hosts for stem rusts that attack grains. These include common barberry (*B. vulgaris*) and American barberry (*B. canadensis*) as well as hybrids resulting from crossing Japanese and common barberries. So if you live in an area where wheat is grown, check with your local Cooperative Extension Service

in the United States or your agricultural authority if you live elsewhere to ascertain which barberries are right for you.

The American colonists used barberry roots (probably those of *B. canadensis*) to make a yellow dye, boiling up great batches of it. Another species of barberry, most likely *B. vulgaris*, was used as a substitute for pomegranate in cookery in the fifteenth century. It was also used in the Middle Ages to season and garnish meat. It has a tingling acidic taste, vaguely reminiscent of citrus. According to Mrs. Beeton's *The Book of Household Management* (1861), the barberry, when boiled with sugar, "makes a very agreeable preserve or jelly. . . ." She also recommends intertwining bunches of barberry with curled parsley, as it makes "an exceedingly pretty garnish for supper-dishes, particularly for white meats."

Barberry can also be used decoratively in crafts, but wear gloves when gathering the branches to protect your hands from its thorns. The berries are particularly well suited to wreaths, hanging like red teardrops among foliage.

Barberry is particularly well suited to wreaths, above, as its thorns help lock it into place. This wreath was created by wiring bunches of sweet annie (Artemisia annua) onto a metal base and then wrapping the barberry into place.

Barberry cheers and brightens the garden even when it's too chilly for the hardiest flowers, opposite page.

BAYBERRY
Myrica pensylvanica

Most people know the bayberry through the enchanting scent of candles and soaps made from the plant's berries. But seeing this compact shrub growing with abandon in the wild, creating a ghostly-looking mantle of silver on the ground, will allow you to appreciate bayberry anew. In autumn, the plants are covered with gray masses of hard berries (sometimes referred to as *nutlets*). The berries are somewhat downy and coated with a white waxy substance used for making dipped candles; this characteristic has inspired the colloquial name *candleberry*.

Break off a cluster of bayberry in the wild and you will detect that universally appreciated aroma; it seems to embody the untamed beauty of its surroundings. Even the leaves of bayberry have a pleasant fragrance. Although the foliage is deciduous, it will cling to the plant right through the beginning of winter.

Breathing in the spicy scent of this plant, it's easy to understand why North American Indian tribes ascribed so much power to it. The Creeks and Seminoles of the eastern woodlands area used it as both a talisman and a medicine, believing the bayberry could exorcise spirits of the departed as well as prevent disease.

Bayberry thrives in dry, sandy soils. If you tend a garden by the sea, you would do very well to add bayberry to your plant collection. (Specify the *Myrica* genus, as the plant *Pimenta acris*, the similarly aromatic bay rum tree, is sometimes also known as bayberry.) It is quite resistant to harsh winds, salt spray, and salt fog. You are likely to find bayberry growing wild rooted on bluffs overlooking the Atlantic. It grows prolifically in masses as high as nine feet. Its range is from Nova Scotia to Florida, but this North American native is also known to grow inland as far west as the Great Lakes area.

Because it reaches its peak in terms of color and fruiting from autumn through winter, bayberry is most charming used in decorations that mirror and evoke the muted colors of the colder months. You can create such an effect by mixing bayberry with everlastings. Recommended dried plants to mix with bayberry include German statice, artemisia, dahlias, and peppergrass—their soft grays and purples harmonize with the silveriness of the berries. Bayberry spikes also look dramatic in spare, contemporary arrangements.

When you harvest the berry clusters for wreaths or arrangements, try adding the plant's leaves to fresh-flower arrangements as well; the oil from the leaves is a wonderful preservative and extends the lives of the fresh flowers by preserving their stems. You also can use the leaves in potpourris.

Bayberry brings a surreal silvery beauty to the landscape. Here it grows on Cape Cod, where locals use the berries for candles and the leaves for potpourris.

BEARBERRY
Arctostaphylos uva-ursi

In the heathlands next to the rolling windblown dunes of Cape Cod, acres upon acres of bearberry hug the ground. This scarlet-berried plant often grows alongside its heath family relatives, the lowbush blueberry and black huckleberry—creating an ecosystem that stabilizes the land while beautifying it.

Bearberry is often mistaken for cranberry by the uninitiated. Not only does it grow in the same environments as the cranberry, but its brilliant red berry also resembles *Vaccinium macrocarpon*. To add to the confusion, writers of the colonial era referred to the cranberry as *bearberry*. Cape Codders often refer to it as *hog cranberry*. Its genus name is actually Greek for "bear's grape," as the black bear is said to be very partial to the hard red berry.

Most people find the taste of the bearberry disagreeable; it has a cottony, mealy taste and is not recommended for consumption. But simply because the bearberry isn't palatable is no reason to dismiss it. It serves admirably as an evergreen ground cover. Bearberry is quite hardy and trails in a vinelike fashion over the ground. It has dear little glossy green leaves, shaped like petite paddles, bunched along the stem. From May through July, white or pink urn-shaped flowers create wildflower meadows from Canada down the Atlantic seaboard to New England, areas to which it is native; it is also a denizen of Europe and northern Asia.

By the end of September, the carmine berries are ready to be devoured by game birds. From late autumn and well into winter, this obliging, showy plant puts on a display of bronze foliage.

The colonists were so enthralled by bearberry's beautiful leaves that they made an astringent from them and even steeped them in tea to combat dysentery. The Algonquian Indians enjoyed mixing the leaves with tobacco and having a good smoke. They called the mixture *kinnikinnick*, yet another common name used to identify bearberry today. In Scandinavia, the leaves are prized as a tanning agent for leather.

Gardeners who have been disappointed by the poor performance of most plants on their property because of sandy soil take delight in bearberry. It thrives under conditions that would be anathema to other plants, growing happily on rocky sites and spreading out beneath trees.

For those who wish to take advantage of the bearberry's beauty, a range of decorations—wreaths, garlands, or flower arrangements—are effective means to show them off. You can even put the lovely foliage to use. In Victorian times, the pressed plant picture came into vogue, along with such other curiosities as hair paintings and carnivorous plant collections. Artistically inclined women created albums filled with easily preserved flowers, like clover, daisies, and baby's breath, as well as inter-

esting greens. The bearberry leaf was one of the first types of foliage pressed in these albums. A spectacular early example is a design by the self-taught Scottish botanist Robert Dick. Although Dick arranged the foliage specimens in the name of science, he also had an artistic eye and created lovely flowing patterns with the plant's ovate leaves. To do so, dry the leaves in a large, heavy book. The process will take about two weeks, although it can sometimes take longer. When the leaves are flat and dry, place them on parchment paper or some other attractive background and cover with a frame. Whether your interest is botanical or purely aesthetic, arranging bearberry leaves is a restful pastime on a cool autumn day.

The first time you encounter bearberry, you might mistake it for the cranberry, a heath family relative. But this plant is valued for the beauty of its fruits and shapes of its leaves rather than for its edibility.

BITTERSWEET

ORIENTAL BITTERSWEET
Celastrus orbiculatus

AMERICAN BITTERSWEET
Celastrus scandens

Take a walk through the woods in summer in eastern North America and you might notice small, unassuming yellow blossoms growing on vines that twine at your feet and climb high up the trunks of trees. Return in September and you will notice that smooth yellow-orange fruits have replaced the blossoms. Stroll through again at the end of the month and you will be startled. The yellow shells of the vines will have burst open, exposing crimson berries within.

So go the seasonal costume changes of bittersweet. American bittersweet, *Celastrus scandens*, is a perennial vine native to eastern North America. This hardy plant is often seen along roadsides and in the woods and can grow to as long as twenty feet, twining around—and ultimately strangling—everything that comes in its path, even curtailing the trunk growth of trees with its thin yet tenacious woody twigs. Bittersweet's beautiful appearance but invasive personality may very well have inspired its name.

In this more naturalistic design, bitter-sweet mixes freely with evergreen foliage in a large basket.

American bittersweet is sometimes referred to as *false bittersweet, shrubby bittersweet,* or even *climbing orangeroot,* in order to distinguish it from another plant also known as bittersweet—*Solanum dulcamara,* a woody vine in the nightshade family whose red berries, twigs, and stems contain a narcotic poison. Of course, you won't be indulging in the plant parts of any of the bittersweets, but be forewarned of the particular danger of this plant.

Similar in appearance and habits to American bittersweet is Oriental bittersweet, *C. orbiculatus,* which is native to Japan and China but has been naturalized through the North American East. The principal difference between the plants is that American bittersweet's berries are borne along the tips of the twigs, while the Oriental species' berries grow along the twigs on lateral offshoots. Also, Oriental bittersweet can grow to nearly twice the length of its smaller American cousin.

Don't let the spreading, twining habits of the *Celastrus* vines deter you from growing them. They can easily be kept in check through being trained to tumble over walls and wooden structures. Some gardeners erect pretty pergola walkways for bittersweet to clamber over. One ingenious gardener I know combines a love of tennis and bittersweet by growing the vine along the metal fence surrounding the court.

Harvest American and Oriental bittersweet after the vines have shed their leaves. This saves you the task of removing them yourself. It is also preferable to pick berry stems before the shells have opened. Otherwise, the berries will be too fragile and drop off the twigs in the drying process. Don't wait after harvesting too long to use your bittersweet decoratively; it is most pliant and most apt to hold on to its berries if arranged just after being picked.

Bittersweet looks altogether beguiling whisking through arrangements of fresh foliage—such as spruce or asparagus greens (*Asparagus sprengeri* is especially nice). Fashioned into wreaths, bittersweet takes on a rustic air, but its personality can also

wax sophisticated when contrasted with fresh flowers in arrangements. Bittersweet is so striking that it doesn't even have to be "arranged"; simply mass bittersweet in a large container for a stunning decorative accent or let it twist and turn along a mantel. A gesture of hospitality is to hang bittersweet sprays on your front door, tied with a bright red satin ribbon.

Bittersweet can be a very versatile home decoration. Here it has been gently coaxed into a miniature heart-shaped wreath. Also try hanging it on your front door, twining it around banisters, or placing it on a manteltop.

Growing on the banks of the Hudson River, bittersweet drinks in moisture. Consider training bittersweet on a wall or trellis, but be careful not to let it spread and twine around plants growing nearby.

176

DIANELLA
Dianella tasmanica

*T*he romantic garden is a place of magical plants, sweet smells, and picturesque enclaves. A bed of dazzling roses, a small rustic bench for intimate conversation, a pergola covered with morning glories, a summerhouse—these create the atmosphere of romance. A berry-bearing plant with the appropriately romantic and lyrical name for such a garden is dianella.

Like marsupials and kiwi, dianella is a native species of Australia. Originating on the island of Tasmania, just off the Australian coast, dianella belongs to the lily family and is sometimes known as *flax lily*. After the pale blue flowers bloom, they are followed by berries, for which the plant is principally grown. Dianella's berries are extraordinary. The glossy fruits nodding on the tips of stems range in color from sapphire to rich purple. The berries are so beautiful that they have inspired collectors to go to great lengths to acquire the plant.

The English gardener and writer Margery Fish, whose horticultural ruminations and wisdom helped inspire the cottage garden craze, was somewhat addicted to planting ornamental berries. She used them prolifically in the glorious jumble of plantings at her homestead, East Lambrook Manor in South Petherton, Somerset, and she wrote enthusiastically in *Gardening in the Shade*

that dianella makes a "wonderful accent plant for a woodland position." As Fish advised, the plants "are safe outside only in the southern half of England and elsewhere are more usually grown in a cold greenhouse." The same gardening strategy holds true for North America.

If dianella has struck your fancy, opt for a sheltered position when you plant it. At Forde Abbey in Somerset, the plant is located in an ideal spot—a walled kitchen garden, which protects it from winds. The berries are so appealing in their natural state that it seems a shame to pluck the beautiful berries from the plants for decorations. So leave them be and enjoy a little Tasmanian romance.

The Forde Abbey in Somerset, England, features dianella in the kitchen garden. The plant's distinctive manner of growing is in keeping with the grand abbey buildings.

So petite and perfect, dianella is guaranteed to charm in the garden.

FIRETHORN
Pyracantha coccinea

Firethorn is the bane of mischievous children, who often fall victim to its clinging thorns when scrambling through the garden. Adults, however, have learned to enjoy firethorn from a distance and to take delight in this berried shrub. After its white flowers blossom in May and June, the shiny, fiery red-orange berries follow and often persist into winter, earning the plant its other common name, *everlasting thorn*. Some firethorn cultivars and varieties have orange and yellow berries as well.

Native to Eurasia, firethorn is an evergreen shrub that can grow fifteen feet high. Its glossy, thick green leaves contrast strikingly with the berries. The plant's genus name, *Pyracantha*, is Greek for "fire" and "thorn," descriptive of the shrub's berries and twigs, respectively. Sometimes known in the past as *Cotoneaster pyracantha*, firethorn is closely related to the *Cotoneaster* genus and requires the same general care:

sunny locations and partially dry to well-drained alkaline soil.

In her book *Edible and Useful Plants of California*, Charlotte Bringle Clarke writes that *P. coccinea*'s berries were once used, presumably by the first California settlers, for jellies and also added to citrus juices. However, the berry's edibility is suspect, and there are more safe and satisfying uses.

You can grow firethorn in a container and move the plant about to suit your fancy. It can be trained against a wall—try espalier fashion—or clipped into a hedge. The shrub is a common sight along roadsides of California, often bordering cow pastures, its thorns acting as an obstacle to otherwise wayward cattle.

When used decoratively, firethorn berries add blazing color around the home. The berry clusters are best suited to rustic, textural arrangements and combine beautifully with twigs, branches, and foliage in wreaths and arrangements.

Firethorn is a familiar sight in California, where it is often planted to create a living fence along pastures. Be forewarned by this shrub's name and wear gloves when you harvest the branches. If you are schooled in bonsai, consider training the plant into an expressive, sculptural form.

A wreath ablaze with firethorn adorns a barn door. The boldly colored berries look best paired with unusual, uncontrived elements, such as oak twigs and eucalyptus bark, which twists through the wreath. Strawflowers, rosehips, and statice used sparingly add texture and subtle color.

HOLLY

ENGLISH HOLLY
Ilex aquifolium

AMERICAN HOLLY
Ilex opaca

WINTERBERRY
Ilex verticillata

Think of holly and images of the snowy landscapes adorned with crisp, shiny green leaves and brilliant scarlet berries come to mind. This wonderful plant also conjures cozy Dickensian scenes of families trimming Christmas trees in festive rooms ornamented with holly garlands and mistletoe. A plant that cheers the garden in the chill of winter, holly's merits have been recognized for centuries, and it has long been associated with the holiday season.

In England and France it used to be a tradition during the first few days of Lent to create plant figures dubbed the Holly Boy and the Ivy Girl and then burn them in effigy. It has been hypothesized, quite fittingly, that this act would signify the end of the holiday season. What people commonly think of as Christian ceremonies actually have their roots in pagan practices. Holly, ivy, mistletoe, and other plants were used as talismans against evil by ancient peoples such as the Celts and the Norsemen. In the early days of Christianity, in fact, the custom of decorating places of worship with traditional pagan greens was banned by church elders because of the links to heathenism. But somewhere along the way they relented, for the customs resurged and halls have been decked with holly for centuries.

Holly has also gained a reputation in areas other than holiday decoration. It has been used in Europe as a divining plant in matters of love. A British folk belief is that if

Ilex pernyi, above, is not as well known as most hollies, but has much to recommend it. The plant, a native of Central China, has endearing little leaves and petite clusters of orange berries.

American holly, *Ilex opaca*, left, can grow to a stately fifty feet on your property, as has this specimen grown at the Brooklyn Botanic Garden and photographed on a chilly December morning.

a holly bears many fruits, a cold winter lies ahead. In Surrey, the holly was used much in the same way as the blackberry was for curing childhood ruptures or rickets, by passing the afflicted youth through the plant headfirst. Witches also fear holly, the old English belief goes, because it was the original component of the Christian crown of thorns, the bright red berries having sprung forth to commemorate the Holy blood of Christ.

Ilex aquifolium is the premier European holly species and the holly tree of Norse mythology, which, in part, gave rise to our use of it as a holiday decoration. It is an evergreen plant that can grow to a towering fifty feet, and its leaves are as well known as the berries. They are thick, glossy, and somewhat elliptical. English holly's foliage inspired the colloquial name *crocodile* in Britain because of its spiny crenellation.

In the United States, English holly thrives in areas with a moist climate free of harsh winters, such as the Pacific Northwest. Many cultivars of this plant are offered, usually with some variation in foliage, such as silver- or golden-rimmed leaves. Sometimes the leaves are even spinier than the traditional forms of *I. aquifolium*, so do be aware of this.

Not all hollies are evergreens. *I. verticillata*, known commonly as *winterberry*, sheds its leaves, but the bright red berries that remain contrast all the more cheerily with their surroundings, particularly on blustery, snowy days. Winterberry, an American native, can grow as high as fifteen feet and is commonly found from Newfoundland down through northeastern states. Indians made a tea from winterberry bark to treat fever.

Another American native is *I. opaca*, American holly, a stately, conically shaped evergreen tree that can grow to fifty feet high. It is prolific in the North American wild. This plant's berries are not as intensely red as those of English holly, and the leaves are generally not as glossy. However, the tree is hardier than its British cousin, and many cultivars have been de-

The Japanese holly, I. serrata, above, is a diminutive yet striking plant. It doesn't mind the cold and looks very pretty well into the blustery days of winter, as proven by its appearance here at the Brooklyn Botanic Garden.

Winterberry, right, is one of the finest hollies to use for decorations because its stems are spiky and strong and its berries grow profusely. It is mixed with foliage in this interior arrangement in a home where it also brightens the grounds with color.

veloped from it that have interesting traits.

Particularly attractive are hollies with petite parts. With its dainty red berries and frondlike leaves, the Japanese holly, *I. serrata*, looks like a miniature version of winterberry. Another small charmer is *I. pernyi*, an evergreen holly native to central China. Its diminutive leaves and clusters of orange-red berries give this holly a coquettish appearance.

Despite the fact that they are often associated with snowy landscapes, evergreen hollies grow best in areas where the winters are only moderately cold. Deciduous hollies are the true cold-weather performers. It is only because of the intensive hybridizing work of Kathleen K. Meserve (to whom the botanical species *I. meserveae* owes its name) that hybrid evergreens that can withstand harsh winters have been developed. Aside from *I. meserveae* hybrids such as 'Blue Boy' and 'Blue Girl', there are also *I. cornuta* hybrids developed by Meserve that are able to withstand the staggering temperature of minus twenty degrees Fahrenheit.

Hollies like fertile and well-drained soil and enjoy full sun or partial shade. Use evergreen holly in the landscape as a hedge or screening plant. Hedge plantings, of course, require extra pruning to encourage a denser side growth. A spring freeze that takes place after the tender young holly berries first appear could cause them to drop. Lack of productivity could also be the result of not having a male and female species—a must for cross-pollination in hollies.

If you wish to accomplish two tasks in one, prune hollies in late December so that the clippings can be used for holiday decorations. Take extra time to shape and prune the plants. Don't recklessly cut off branches for use as decorations, however. If you're tempted to take a long berry-bedecked branch, be sure that it won't mar the overall appearance of the plant. The least desirable time to obtain cuttings for arrangements is when the foliage seems droopy (the plant is already ailing and the

greens will not last long in an arrangement made with these).

After you have gathered holly for decorations, it's wise to condition it to ensure long-lasting designs. They should also be thoroughly—yet gently—cleaned. Submerge them in cold water for a while and whisk them around to remove dirt. Do not use metal containers for the process, as they rust. Leave the holly in the water for two to three hours. Afterward, split the cut stems at the base to ensure good water flow and place the holly in water overnight or until you are ready to use it.

You can put holly to use decoratively in many ways, and it keeps for a long time. It is so elegant that it can simply be laid on a mantelpiece or table. Other possibilities include wreaths, garlands, swags, and container arrangements, either by themselves or combined with evergreens. One wreath designer advises that you wear gloves when working with holly, as the prickliness has been known to cause allergic reactions in some people and the spiny leaves just plain pain. To give yourself more leisure time during the harried holiday season, she suggests that you make your holly decorations a few weeks in advance and store them in a cool spot. For her holly wreaths, she simply gathers small bunches and wires them onto a wire base with one continuous piece of 26-gauge wire. The holly almost arranges itself, its foliage providing texture and form and its berries acting as natural focal points in the decorations.

English holly is the best loved species of all. Its bright berries and defined, glossy leaves have inspired many folk beliefs.

For long-lasting beauty, holly is a natural choice for plant decorations. This wreath, left, will stay fresh looking for months.

This deciduous holly thrives at the Brooklyn Botanic Garden, opposite page.

189

PEPPERBERRY
Schinus molle

Garden historians have speculated that the peppertree once lined the avenues of Cuzco, Peru, capital of the Incan empire. With its flowing, weeping form, the peppertree caught the fancy of the Spanish conquistadors as well. The New World Spanish missionaries of the sixteenth century so prized the native South American tree that they imported it to California to provide shade and beauty on the grounds of their missions. The tree is so prolific now in California, it is sometimes known as *California peppertree.*

The peppertree's gorgeous rose-colored berries emerge in summer and last through winter. The tree much resembles a weeping willow in form and foliage, and the beautiful evergreen grows to as high as fifty feet. Despite its name, the tree has no relation to the spice, and the berries are inedible. However, the berries have much to recommend them for decorative use.

Come late summer, pepperberries, borne profusely in long clusters, are a common sight at flower stands all over the United States. In such cases the berries have been dried and there is no foliage on the plant. However, if you are a California resident or live in the American South—where the peppertree's close relative *S. terebinthifolius,* known as the *Brazilian peppertree* and *Christmasberry tree,*

thrives—you might try harvesting your own pepperberries. When the berries have deepened to a rose color, cut the stems and strip the leaves.

A northern California floral designer who works extensively with pepperberries, recommends drying the berries by hanging the stems in a very warm spot with good ventilation. The more quickly the berries dry without being exposed to moisture, the better they will retain their original color, she advises. She uses pepperberries as accents in wreaths and garlands, working them in with such everlastings as roses, bells of Ireland, spiky red leptospermum, echinops, wheat celosia, eucalyptus, ammobiums, and irises.

Pepperberries are also stunning when used in floral arrangements. Another northern California floral designer uses pepperberries as dramatic accents. She harvests fresh racemes of berries—with foliage intact—and weaves them through designs that incorporate fresh flowers and foliage, such as roses, delphiniums, and eucalyptus. One of her favorite ways to use the berries is to mass them as focal points in arrangements and allow them to trail over the sides—a very dramatic, Victorian style that looks particularly lovely on a sideboard during the holiday season.

The graceful pepperberry tree much resembles the weeping willow, although the two are not related. The pepperberry tree is actually in the same family as Rhus, or sumac. Both bear distinctively colored, small, hard berries.

190

Massed pepperberries spill over the sides of this arrangement evocative of the Victorian age, left. The design also contains an abundance of freshly cut roses, liatris, alstroemeria, and delphiniums.

You can dry pepperberries by hanging them upside down for several weeks. Some floral designers remove the foliage before drying the berries.

ROSEHIPS

BURNET ROSE
Rosa pimpinellifolia (spinosissima)

RUGOSA ROSE
Rosa rugosa

CALIFORNIA ROSE
Rosa californica

For many centuries, gardeners have adored the gorgeous aromatic flowers of the rose, but they have also prized the hips. These are the oval or rounded fruits of the rose that appear in late summer and autumn. A rosehip is actually a receptacle that encloses the true fruits of the plant, the achenes, or "seeds." Some linguists contend that the original English word for rose was *hip*.

Because the rose is a sacred flower, its hips have been naturally accorded the same respect. In his late-nineteenth-century work *Flower Lore*, Hilderic Friend speculated that the Catholic rosary was so named because rosehips were once used to count the prayers as they were said. The beads of the rosary do indeed resemble smooth, elongated rosehips, similar to the graceful, long-necked fruits of the Chinese species *R. sweginzowii*.

Other species lauded for their hips include the rugosa rose (*R. rugosa*) and *R. villosa* (*R. pomifera*). The rugosa, or Turkestan rose, which originated in the Orient, is laden in autumn with plump, brilliant red fruits and is considered by hip hunters to be the most flavorful plant for eating fresh or cooked. It is an upright thorny bush with big, fragrant reddish-pink flowers with a long blooming season. This shrub especially enjoys living near the seashore. Its dense form and deep green foliage make it an excellent candidate for hedge plantings. The pink-flowered *R. villosa* is native from Europe to Iran and its fruit is particularly pleasant when made into a delicious juice.

Barnsley House in Gloucestershire, England, is the backdrop for this impressive planting of the rugosa rose, plump ruby hips in full splendor.

One of the most unusual types of rosehips are those of *R. pimpinellifolia*, the Burnet or Scotch rose. The hips have a sable or purplish coloration. Garden designer Gertrude Jekyll very much favored this native British plant, which still grows wild on sand dunes and in limestone areas. Equally fond of the single flowers of the wild plant and of the double garden varieties, Miss Jekyll wrote enthusiastically about the flowers and foliage and characterized the hips as "large and handsome, black and glossy." The arrival of the hips, she enthused, is complemented by the shrub's changing to "a fine bronzy colouring between ashy black and dusky red."

If this eloquent recommendation still leaves you unimpressed, consider the Scotch rose's history. It was one of the first roses to be studied widely by plant breeders. The experimentation began in 1793 when a nurseryman in Perth, Scotland, transplanted the species from the wild into his experimental plot. He was interested in producing variations on the naturally creamy-white flowers and with each successive experiment was able to redden slightly the petals of the shrub and to alter the structure of the petals to create semidouble flowers, all with a characteristically delicious, sweet scent. A half century later, more than three hundred varieties of this humble wild rose existed.

You have many options for using roses in the landscape. You may wish simply to create scattered hedge plantings or perhaps incorporate borders of roses into the kitchen garden. Creating a special rose garden is a splendid idea not only because of the incredible seasonal displays of massed summer flowers and autumn hips that will result, but for purely horticultural reasons as well. Roses are hungry and aggressive plants and tend to compete with surrounding plants for nutrients and water in the soil. Also, roses need to be able to spread out, as overcrowding may result in plant diseases. In general, species roses are easier to care for than hybrids, and they also tend to produce more succulent hips

for eating, so if your intention is culinary as well as ornamental, you may wish to cultivate plants from this group. If you plant climbing roses and want them to fruit, remember, do not prune them directly after their summer flowering.

Rosehips will generally remain on the plant until the birds begin to eat them—during November or December—or until you decide to harvest them for your own larder. Another possible rosehip robber is the mouse. Look into a field-mouse hole in winter and you're likely to find it filled with hundreds of discarded rosehip seeds.

The hips, which have a zesty acidic but fruity taste, are the most nutritious part of the plant. Fresh rosehips are said to contain sixty times as much vitamin C as oranges, and rugosa roses, with their large round fruits, are considered to have one of the highest contents. You can eat them raw, but eat only the walls of the hips and discard the seeds.

Rosehip syrup has long been popular in England, believed valuable in keeping young Britons healthy. A British gardener said wistfully while admiring his hedge of rugosa roses, "When I was a child, every

Topiaries should have an element of fantasy. This decoration, left, pairs rosehips with gilded leaves.

Spiny-hipped roses, opposite page, are not desirable for eating, but make novel garden accents. These barbed beauties grow at Tintinhull House in Somerset, England.

197

Regal rosehips are excellent choices for topiaries. During the holiday season, tabletop topiaries fashioned into conical evergreen shapes make pretty and appropriate decorations, right. This topiary combines dainty _Rosa multiflora_ hips, tansy, and sweet annie _(Artemesia)_.

The lustrous ebony hips of the Burnet rose, opposite page, have long been favorites of autumn garden connoisseurs. This rose is usually spotted growing in the wild near seaside bluffs.

day I would have a spoonful of rosehip syrup, and it kept you healthy." Rosehip syrup was a staple in British households during World War II in particular, when it was difficult to import oranges.

If you are curious, here is an age-old recipe for that vitamin-packed rosehip syrup, from a resident of Somerset, England. She writes, "Use approximately 2 pounds of hips for every 6 pints of water. Mince the hips after removing the stalk and calyces. Cover with water and boil, then strain through a jelly bag. Reduce liquid to half, add 1 pound of sugar, and boil for 5 minutes to sterilize. Pour mixture into sterilized bottles and seal up right away with sterilized stoppers."

Rosehips, by association with the much-valued rose, have a certain cachet and have eagerly been eaten for centuries by the upper crust. In his sixteenth-century book *The Herball*, John Gerard attested to the rosehip's gourmet status: "The fruit when it is ripe maketh most pleasant meats and banqueting dishes, as tarts and such like; the making whereof I commit to the cunning cooke, and teeth to eate them in the rich mans [sic] mouth."

Don't think of rosehips simply as "health food." You can also make them into preserves, sauces, juices, soups, teas, and wine—or you might even try the old-fashioned method of baking them into tarts, as Gerard suggested. Such a recipe was committed to paper by American colonist Robert May in 1671. Of course, the directions are sketchy but should provide enough clues for enterprising cooks to experiment with. They call for seasoning the hips with "sugar, cinnamon, and ginger." Then you must "close the tart, bake it, scrape on sugar, and serve it in."

While cooks discard the seeds covered in bristly hairs, British children long ago found a way to put them to use. The same gardener who used to drink rosehip syrup daily also reminisced about a fiendish schoolboy trick. He and his friends used to bring the hairy seeds to school and slip them down classmates' shirts. "A few of those down your back and you'll be itching all day," he cackled.

Most of us will limit our use of the rose to ornamentation and cooking. When using roses for cooking, don't forget about the petal potential. Fresh rose petals add color and drama to salads and are also elegant touches in omelets. Wines, juices, liqueurs, teas, jams, and fragrant waters have also been made from them. Rugosa rose petals are particularly strong-tasting. A Britisher newly arrived in the American colonies once instructed his son to bring only a few vital provisions, among them peas, figs, sugar, oatmeal, and, of course, "conserve of redd roses," which could have been made from rosehips or petals. And in the 1653 publication *A Book of Fruits and Flowers*, with no author listed but commissioned by the bookseller and print dealer Thomas Jenner, there is a recipe for a syrup of damask roses, guaranteed to purge melancholy.

Once you think you have exhausted uses for roses, you will always find something new and delightful you can do with them. A stunning option is to use rosehips decoratively indoors. You can add them to wreaths, put them into fragrant potpourris, or even use them in topiaries. A clever way to defy the seasons is to create a wreath of dried miniature roses and hips, combining the colorful flowers of summer with the delectable fruits of autumn.

Rosehip tea, left, is a soothing complement to preserves and teacakes.

You don't need to harvest rosehips for food or decoration to have them serve a purpose. They are also spectacular looking in the garden, right.

SAINT-JOHN'S-WORT
Hypericum androsaemum

Understated in appearance, the low-growing black-berried Saint-John's-wort does not boldly proclaim its rich history. It is a plant that has long been thought to have great powers. Peasants of the British Isles summed up the powers of the herb neatly in the rhyme "Trefoil, Vervain, John's-wort, Dill,/Hinder witches of their will."

The plant was once called *Fuga daemonum*, translating roughly as "Flee demons!" Hilderic Friend, in his Victorian-era opus *Flower Lore*, ventured to assert as evidence of its magical associations that the name *Hypericum* is derived from a Greek word meaning "to hold over in such a way as to protect from anything." Less adventurous horticultural tomes simply state that *Hypericum* is an old Greek plant name possibly meaning "beneath or among the heather," because of the low-growing characteristic of many of the undershrubs that comprise the genus.

European lore has it that anyone who picks Saint-John's-wort on June 23, Saint John's Eve (also called Midsummer Eve), will have the power to see witches holding their traditional yearly festivities on Saint John's Day (Midsummer Day)—and consequently be able to avoid them. People commonly climbed to the roofs of their homes to get a bird's-eye view into the landscape to spot the hags. Perhaps this is what gave rise to the custom of placing wreaths fashioned from Saint-John's-wort on the rooftops of homes as a general protection against evil.

The purported powers of Saint-John's-wort are not limited to its namesake's day. In many European regions it was commonly planted by the door, hung up in the house, or burned in midsummer fires to ensure protection from the dark forces.

Physicians have long believed that Saint-John's-wort also provides protection of a corporeal nature. In ancient Greece and Rome, doctors used the leaves of some species to clean wounds. Even up to the eighteenth century the plant was being used to treat skin problems as well as nervous disorders. By the Victorian era, it was no longer employed so liberally for these maladies. But as Saint-John's-wort is a tenacious herb, interest in it has been reawakened and parts of the plant are being used in an experimental fashion to treat nervous complaints. Be aware that the plant does have toxic properties, so do not devise any home remedies using the plant.

With so much intrigue surrounding Saint-John's-wort, surely it deserves a place in your garden. Many a gardener has been charmed by this herb, which is easily cultivated and can grow in partial shade. At home in the rock garden or the border, perhaps planted in the foreground of a stone wall, this unassuming undershrub is a joy to have in the garden. In midsummer its yellow flowers perk up the garden. Its natural habitat is woodlands and meadows. *H. androsaemum* is native to Europe and western Asia; *H. perforatum* is also native to Europe but has been naturalized in North America.

Its foliage spreads thickly on the ground, prompting fussbudgety gardeners to dub it a "weed"; however, if it is kept in check, it grows very nicely and agreeably. But it is the berries—known as *capsules* in botanical terms—that make Saint-John's-wort such a cherished plant. They can be a nice counterpoint to the tiny flowers and foliage of dwarf sedums and other low-growing rock garden plants. You can also use the prized berries for autumn decorations, as they create a rustic effect paired with flowers and greens in vases or wreaths.

The berries of Saint-John's-wort turn from red to black as they mature. Here it grows in a cottage-garden-style ensemble with lobelia and <u>Sedum spectabilis</u>.

SNOWBERRY

SNOWBERRY
Symphoricarpos albus

CORALBERRY, INDIAN CURRANT
Symphoricarpos orbiculatus

With its ghostly-white, winter-persistent fruits, the snowberry could have been so named for both its appearance and the likelihood of seeing the fruits clinging to the branch even when it's flurrying outside. Most Americans are unfamiliar with this group of showy pink- or white-flowered shrubs—great favorites of bees and hummingbirds—even though of the total of sixteen *Symphoricarpos* species, only one is not indigenous to North America. The one most frequently seen growing in American gardens is the alabaster-berried *S. albus*, which has lovely marble-sized berries that emerge as early as midsummer.

In England, these robust plants are more widely appreciated than they are in the United States and are regular features of London gardens because of the genus's resistance to pollution. The snowberries also respond well to a variety of soils. Moreover, although they certainly enjoy full sun, they will send forth their plump alabaster fruits even in the shade. Margery

Symphoricarpos plants, left, are featured prominently in the borders of this garden in Somerset, England.

Fish attested to this in *Gardening in the Shade* when she wrote, "They do extremely well in the most uninviting places under trees and as birds do not care for the berries they look as attractive in the winter garden as they do in flower arrangements in the home." Few plants give the gardener so much pleasure in exchange for so little toil in the earth.

Snowberries are dense shrubs that make excellent hedges, although they also make eye-catching specimen plantings. They are also attractive when used for covering slopes. The elaborate genus name of the plants, *Symphoricarpos*, comes from the Greek for "bearing together" and "fruit," descriptive of the striking clusters of globular berries that emerge in summer.

Another *Symphoricarpos* species, *S. orbiculatus*, produces berries that are pinkish white mottled with fuchsia. This shrub, known as *coralberry* and Indian currant, also has fiery autumn foliage. Hybrid *Symphoricarpos* species often have interesting, marbleized patterns on the berries or unusual, variegated leaves.

A snowberry plant undergoes a startling transformation from flower to berry at Wisley Garden in England, left.

SUMAC

SCARLET SUMAC
Rhus glabra

STAGHORN SUMAC
Rhus typhina

*E*ven after the last of the colorful autumn leaves drop, majestic minarets of scarlet sumac berries blaze in the eastern part of the United States. At this time of year, scarlet sumac brings its glow to shrub borders and city gardens on both sides of the Atlantic.

Some gardeners covet this woody plant for its pinnate foliage, which turns yellow in autumn, as well as for its graceful form, but scarlet sumac also excels in its fruit, red berries densely grouped in striking rhomboids. It can grow to twenty feet—not quite as tall as its close relative *Rhus typhina*, staghorn sumac, which bears similar berries. Both species are natives of temperate eastern North America and are hardy plants, able to withstand frost and pollution. They are lovely enough to use as specimen plantings in the garden, but you can also grow them on your city terrace in any oversized containers.

Some people shy away from sumac because of the well-known toxic properties of other members of the genus, the poison sumacs. The sumacs *RR. radicans*, *diversiloba*, *toxicodendron*, and *vernix*—now also grouped under the genus name *Toxicodendron*—all are contact poisons, but their berries are white or gray (*vernix*'s is greenish-white), so it would be impossible to mistake them for scarlet sumac's red berries. If there is any doubt, have the plant identified by a local authority, such as a botanical garden.

Although scarlet sumac's most obvious use is ornamental, it has been used for other purposes. It has a cult following among natural-food enthusiasts, who make a lemonadelike drink from the berries by steeping them in water and adding sugar, similar to an old American Indian practice. The American colonists preferred to capitalize on the berries for their color,

boiling them to obtain an earthy red dye. The branches, boiled along with the berries, yield a strong, indelible black ink that reputedly grows darker with each washing when applied to cloth. Because of the ink's long-lasting properties, children of colonial times used it to mark their names on objects, an early form of graffiti. Another Early American use for scarlet sumac berries was as an astringent; they were mixed with some liquid, presumably water. The leaves were applied as a poultice to soothe the burn of poison ivy. Mountain people have been known to simmer the berries of scarlet sumac to make a gargle.

The less adventurous among us prefer simply to admire the extraordinary color and form of sumac berries. Both the scarlet and the staghorn types can bring a geometric air to flower arrangements when the berry panicles are placed in whole. You can also break them from the plant for use in herb wreaths; the berry clusters add bright spots of color.

White pine foliage, German statice, a sumac spire, and viburnum berries are the simple ingredients for this stunning and long-lasting arrangement, left. All of these plants — or plants similar to them in appearance — are easily obtained on a leisurely stroll through the woods in autumn.

Sumac harmonizes with autumn wild-flowers and reeds in this spare, Japanese-influenced arrangement, right. Rhom-boids of sumac berries lend themselves nice-ly to arrangements that require minimal yet strongly shaped elements.

Scarlet sumac berries are striking in the landscape, but they're also used as a fla-voring agent in beverages and as dyes. You don't need to be a country dweller to obtain them: they are often available in stores that sell herbs and spices.

VIBURNUM

TEA VIBURNUM
Viburnum setigerum

LINDEN VIBURNUM
Viburnum dilatatum

*V*iburnums are arguably the most beautiful and bounteous ornamental garden plants. There are more than two hundred species in this immense genus, mostly consisting of deciduous shrubs. In addition to lovely berries that range in color from orange to red to black, many have fragrant spring flowers and striking autumn foliage. If you're a bird-watcher, having viburnums on your property will be particularly gratifying in autumn, winter, and sometimes even spring—if the berries persist on the plant— as blackbirds, waxwings, and other species descend on the berries.

Among the most beautiful of the viburnums is *Viburnum setigerum*, the tea viburnum, a native of China. Some people claim its common name refers to an ancient practice among Chinese monks of making tea from its leaves. It grows to about twelve feet high and in autumn is covered with masses of carmine berries. One cultivar, 'Aurantiacum,' has stunning orange berries. Another gorgeous type is *V. dilatatum*, the

Linden viburnum's cluster of red berries makes a strong impact in the garden border, opposite page. This Japanese species is one of the most popular viburnums.

Resembling bright orange tassels or inverted parachutes, the berries of the tea viburnum cultivar 'Aurantiacum' are cheerful natural ornaments in this frontyard garden, shown at left.

linden viburnum. This Japanese species' bright, shiny berries look temptingly like red-hot cinnamon candies.

There are so many other viburnums to choose from that it's a good idea to browse through a comprehensive horticultural reference book to see which species are particularly well suited to your region. For example, the black-berried Laurustinus viburnum (*V. tinus*), can be cultivated only in very warm places. Other species, such as the nannyberry viburnum (*V. prunifolium*), are most adaptable and can grow from Texas to Connecticut in the United States. Handsome viburnums to consider planting are the American cranberry bush (*V. trilobum*) or its close European relative, the guelder rose (*V. opulus*). Both have striking scarlet winter-persistent fruits. Viburnums make stunning specimen plantings. They are also nice border plants, especially when massed in groups of three or more. There are diminutive varieties that are well suited to small-scale sites and to container plantings.

Viburnum berries are dazzling in arrangements mixed with pine greens, autumn leaves, and other unfussy elements. They also act as focal points, glowing in tiny bunches at different parts of the arrangement. Try adding the lovely berries of linden viburnum to wreaths in lieu of the more common holly berries.

Since the linden viburnum bears berries so profusely, try using a branchful to perk up a wreath of everlastings, above.

Viburnum berries are as compellingly beautiful as the face of the rarest flower — and much longer lived. Although many viburnums have red berries, right, you can also obtain them in orange and black.

Chapter Six

BERRY DECORATIONS:
ARRANGING ORNAMENTAL BERRIES

Since ornamental berries can last all year, it's possible to vary your designs to suit the seasons. For colder months, textured wreaths and arrangements using grasses, pods, and orange or red berries are stunning. On a dark winter day, it's heartening to come home to an arrangement of beautiful flame-colored berries or a wreath of barberry and fragrant sweet annie (*Artemisia annua*) hung before the hearth. A small, spare arrangement of sparkling white snowberries and crimson winterberries is also beautiful, the bunches of plump snowberries acting as a base and the spikes of winterberries adding height and providing a framework for the design. In cold weather, consider stringing dried berries with popcorn and peanuts on a nylon fishing line and hanging them in trees for birds to feast on.

Warm-weather arrangements can feature dainty light red Tartarian honeysuckle berries teamed with any number of flowers from the garden. A range of viburnum species also produces lustrous berries in summer in numerous colors. Or you can combine berries brought in during autumn and winter with the bounty of the flower garden at its height.

Designers who work with ornamental berries emphasize their versatility. When fashioning designs using ornamental berries, discard any preconceived notions. Berries need not be "filler" materials. Like flowers, ornamental berries can evoke different moods, depending on how they are placed in arrangements, so allow yourself the same freedom you would with flowers and unusual greens. A floral designer who often works with a variety of berries says, "When I design, I go with my feelings. I don't put any limits on my designs, especially if I'm doing something creative for someone's home or to please myself. Don't overthink the piece. Get a feeling for the nature of the berries."

Bittersweet vines have naturally beautiful curves that can create movement in a floral design. In this arrangement, the dynamic twists of bittersweet contrast with the serenity of peach-tinged white roses. A container woven from palm fibers provides further textural contrast.

STYLES

One of the first things you must determine when creating berry decorations is whether you want berries to play a starring or supporting role. It really depends on the nature of the design. One designer who runs a florist's shop advises that when you are working with striking materials such as bittersweet, you should use them to full dramatic advantage. Accordingly, she tends to mass berries together in flower arrangements rather than space them evenly throughout. This technique creates a focal point for the design and makes for an unusual and strong visual impact.

Some ornamental berries work equally well as bold and subtle touches in designs. Pepperberry's tiny rose-colored berries are dazzling when grouped in an arrangement, but when the berries are used individually in a garland, they stand out as more subtle highlights.

When designing with berries in arrangements, try to establish a basic form, just as you would with flowers and other natural materials. This can be a symmetrical triangle, which evokes a Victorian mood, especially when the arrangement contains many elements spilling out over the sides of the container. The arrangement can also be an oblong form, an asymmetrical triangle, a linear style (characteristic of Oriental designs), or you can choose a traditional rounded form.

The nature of the berries themselves can dictate the style of the arrangement. A stately berried green such as holly combines nicely with other greens in a formal, symmetrically rounded arrangement. A "wilder"-looking berry, such as bittersweet, is well suited to more casual arrangements and can be used twisted throughout the design.

Another floral designer gives the following advice on deciding how to use berries in a design: "First I take a look at the berry branch and see if it has a definite line, shape, or form that would lend itself to a particular style. Look at the habit of growth and see if it's pleasing in itself and determine what you can do to set it off or blend it into another arrangement. For example, you can do something stark in the Japanese manner if you have a dramatic piece."

Consider the effect you want to achieve. Loose arrangements of just a few materials impart a relaxed, welcoming quality.

When using just a few materials in designs, think not only of the interplay of colors and line, but also of the textures of the piece. Contrast smooth petals and berries with feathery foliage or craggy branches. If you prefer working with a wide selection of natural materials, combine colors judiciously and harmonize materials—that is, mix low, rounded elements and taller spiky materials, such as branches. You can even use the berries themselves if they are borne on long stems.

This lovely wreath made in Cape Cod combines locally grown silver-tinged bayberries with everlastings in soft colors.

COLOR

An analogously colored arrangement of flowers and berries suggests autumn.

The principles of color are not as complicated as they appear. There are twelve divisions in a color wheel: yellow, yellow-green, green, blue-green, blue, blue-violet, violet, red-violet, red, red-orange, orange, and yellow-orange. The three primary colors are yellow, blue, and red. The secondary colors —green, violet, and orange—are created from mixing the primary colors. Intermediate colors, derived from mixing primary and secondary colors, are all the in-between colors. For example, red mixed with orange produces red-orange.

You have a wide spectrum of colors to choose from when working with berries. The key to combining them with other materials is understanding very basic principles of color theory. *Complementary colors* face one another on the color wheel. Examples are orange and blue, red and green, and yellow and violet. Complementary combinations, therefore, are orange viburnum berries with blue larkspur, red holly berries and their own foliage, and yellow freesia with violet beautyberries.

Triadic colors refers to three colors spaced evenly apart on the color wheel. Examples of these colors in terms of berries are orange firethorn berries with their own green foliage intact paired with violet-colored wildflowers.

Analogous colors are groups of three or four colors that touch one another on the wheel. Keep these color combinations in mind when creating designs, but by all means vary them with tints and shades of colors. For example, pepperberries— which are light rose in color and therefore are a tint of red—pair beautifully with spiky lavender liatris and blue-violet larkspur in an arrangement based on analogous color theory.

Most designers who work with natural materials are aware of the principles of color theory but also have an innate sense of color and experiment freely with combinations. A professional designer says, "I've studied color theory in courses. I think it's advisable. Some people just have a great sense of color, but even if you have a feeling for it you should know the basics."

Do not underestimate the role of white in berry decorations. White can heighten the fiery color of berries in a design, and berries that are white themselves—such as snowberry and baneberry—can do wonders for arrangements that incorporate sometimes-unexciting-looking everlastings. One California floral designer is a firm believer in the role of white in arrangements. "I always start with white, and I always use white in everything I do because it is a brightener for other colors," she says.

Gather the unusual, transparent berries of mistletoe honeysuckle, Lonicera quinquelocularis, in winter to add a subtle touch of white to arrangements.

BERRIES IN WREATHS, GARLANDS, AND TOPIARIES

*B*ecause they are so striking, berries naturally lend themselves to special decorations such as wreaths, garlands, and topiaries. There are a number of ways to fashion ornamental berry wreaths. For wreaths that contain elements that must be wired on in bunches, most designers prefer working with wire bases. One floral arranger who specializes in designing wreaths creates individual bunches of mixed elements and wires them successively around the base. You can either combine berries with the bunches throughout the wreath or wire them on in bold clusters at the top or bottom of the wreath, taking the place of the more popular, traditional bow.

Grapevine wreaths work nicely with berries with sturdy stems, such as sumac. With these types of wreaths, the base is meant to show, so the materials used should have a rustic, casual feeling about them. Vegetables, such as chile peppers, and herbs look very homespun when combined with berries in a grapevine wreath.

Foam or straw wreaths work well with cranberries and rosehips because you can simply attach them by cutting toothpicks in half and using them to skewer the berries to the wreath. Another option is to use floral glue to attach the berries. If you are mixing larger elements on such wreaths, such as apples, use florist's picks as skewers. You may want to put down a base of ivy leaves or some other foliage, attached with fern pins, to camouflage the base completely.

Garlands are charming, old-fashioned

A traditional eighteenth-century "kissing ball" incorporates mistletoe foliage and holly berries, above.

decorations and can be varied endlessly. They are usually built on wire bases. One California designer simply intertwines two coat hangers to create a base. Then she wires bunches of fresh latifolia statice along the base. After the statice is attached, she bends the base into a half-moon shape and lets the statice dry for a few days. She uses floral glue to add everlastings. Finally, also using glue, she adds berries in select yet eye-catching places around the garland and then attaches bows at either corner.

Topiaries (arrangements that resemble trees or standards) look extravagant and are long-lasting. Even though they are time-consuming to make, the elegant results are worth it. To create a topiary tree, obtain a Styrofoam base in the shape you want and line it with a sheet of English moss, either pinned or glued on the base so that the white doesn't show through. Then add your berries. If the stems are stiff enough, they can be directly poked in. One designer suggests poking the entire piece of Styrofoam with toothpicks to create holes for the berry stems to push through. She finds that rosehips are excellent berries for topiary, because their stems are generally firm. If your berries don't have stiff stems, use glue or a glue gun to anchor them.

A grapevine wreath is a nice backdrop for strongly colored berries. In this design, left, herbs, tiny spears of red sumac, and plump, purple pokeweed berries — which are beautiful but extremely poisonous — are casually arranged on the grapevine base and anchored using only their stems.

COMBINING BERRIES WITH FLOWERS

Freshly cut flowers are available at all times of the year, so it's possible to combine them with berries even during the coldest months. When designing with berries and fresh flowers, the key is to select long-lasting flowers because berries themselves are so long-lived. Among long-lasting flowers to include with berries in arrangements are ranunculus, grape hyacinths, chry-santhemums, orchids, achillea, zinnia, agapanthus, and lily of the valley.

Many wildflowers are also complementary to berries. Once cut, they tend not to last very long, but they do impart a natural, unstudied quality to arrangements. Among wildflowers to consider using are yarrow, cornflowers, oxeye daisy, maiden pink, and prairie coneflower. Other wild plants that work well with berries are dock, wild hop, wild privet, dead nettle, heather, and yellow toad flax.

Of course, all the traditional everlastings—dock, celosia, eucalyptus, clover, and red bottlebrush, among others—are beautiful additions to berry decorations, not only in arrangements but also in wreaths, garlands, topiaries, and centerpieces for dinner parties. Use your imagination for these occasions.

The most long-lived combination of berries and flowers is undoubtedly that of dried berries and everlastings with which you can create designs that can be used as keepsakes. This garland for an anniversary celebration, opposite page and left, blends dried pepperberries and greens with a range of everlastings, including echinops, cornflowers, helipterum, globe amaranth, dried irises, and ammobiums.

SPECIAL TECHNIQUES

Achieving a professional look and truly spectacular-looking berry decorations often requires a few extra preparation steps. For example, you may need to use florist's picks with some of the weaker-stemmed berries if you are attaching them to decorations. These are simple to use. Just align the pick with the stem and wind the wire around both of the pieces. Cover with florist's tape, preferably in a shade that blends with the stem.

It's fairly simple to mix berries and flowers together in a vase. If you want the elements of your design to stay precisely placed, cut Oasis foam with a large kitchen knife or pen knife so that it will fit into the container. Oasis holds several times its weight in water, so it supplies constant moisture to the stems. You may also wish to add floral preservative. If you are using berries with everlastings, simply use Styrofoam as a base.

Berries massed with flowers in baskets are always striking. For everlastings, simply use a knife to contour the Styrofoam to the basket's dimensions, fitting it snugly. However, for fresh flowers you must line the basket with florist's foil or heavyweight plastic and cut Oasis foam to the dimensions of the basket. Make diagonal cuts in the flowers and place them in. You may have to wire the Oasis to the basket or use floral tape to keep it in place. Alternatively, you could create an arrangement in a container filled with Oasis foam and fit this inside the basket.

If you want berries, flowers, and greens to trail over the sides of an arrangement, you can lengthen their stems while ensuring a fresh water supply to them by using florist's plastic water tubes—simply fill them with water, stick the stem through the opening in the rubber top, and attach a wire to the stem. Place the other end of the wire in the Oasis or Styrofoam base in the container. This way, the berries need not be anchored directly in foam.

Some designers recommend preserving berries by applying a high-gloss polyurethane lacquer to the berries to keep them plump and firmly fixed to the branches. This method does tend to work but isn't absolutely necessary. It is most effective for soft, fleshy berries. You can also dry berries by stripping the foliage and hanging them upside down.

Foliage is a staple in berry decorations. You can keep greens fresh longer by submerging them in water overnight before using them. You can also preserve them indefinitely—although they will no longer look freshly picked—by soaking them in a solution of one part glycerine to two parts very warm water for two to three days.

Another technique that professional designers sometimes employ is painting berry decorations. There are floral paint sprays available in many colors, so you have the option of endlessly altering and refining the colors of dried natural materials. One designer particularly favors gold floral paint and thinks its use should not be confined to the holiday season. She says, "When you add a bit of paint to a dried hydrangea or berry, you can heighten what's there and bring fantasy and sparkle to it." Consider employing such unusual effects for berry decorations made for any of your special occasions.

English holly combined with fruits creates a festive holiday atmosphere in this eighteenth-century American dining room.

Chapter Seven

BERRY-PICKING EXPEDITIONS

WHERE TO
FIND BERRIES

Many purists insist on foraging for their berries because they believe that wild varieties taste better than cultivated types. They also argue that the wild berries are free of pesticides and that their nutritional contents are higher than those of cultivated berries, although this point is debatable. If you are eager to experience berry picking in the wild, there are a number of things you must keep in mind before you make your foray into the woods, bogs, meadows, or mountains.

First, you must determine a likely location for finding wild berries. The individual entries in this book should give you an idea of where to look, but generalizations can be made. Bramble berries tend to grow in roadside hedgerows but can also be found in woodlands. Blueberries usually grow wild in northern regions of North America, some in mountains, others in swamps. Like huckleberries, they prefer acid-enriched soil and are likely to spring up on sites recently ravaged by fires. British bilberries also tend to thrive on the acidic soils of heaths and moors.

In England, you might find currants growing wild, escapees from kitchen gardens. Look for these in woodlands by the sides of streams or in hedgerows. Natural sites for cranberries are bogs and damp moorlands. Perhaps you will even be lucky enough to find wild strawberries. In his *Stalking the Wild Asparagus* (1962), Euell Gibbons called these "the top prize for the wild food gatherer." Wild strawberries tend to grow on sunny, open slopes.

Field guides that focus specifically on your region will also be a help in locating and identifying wild berries. If you are ever even the least bit unsure of a berry's identity, do not harvest it or sample it, as it may be a poisonous species. Slowly build an understanding of what plant communities grow in your area. Always bring along gloves to protect your hands from thorns or contact poisons from wild plants. If you are venturing into a swampy area, wear sturdy high boots. If you plan on going on a hike, obtain a good walking guide that discusses the flora and fauna of a particular region. Guides such as these generally list what types of plants you will find growing in different areas.

Also keep in mind that, legally speaking, you must obtain permission from the landowner or whoever administers the land to pick berries on it. The land might be owned by a farmer, or it could be public property. However, most rural people are familiar with at least one or two long-established berry-picking enclaves, where the tradition of berrying far outweighs any legal technicalities, and if you ask around, maybe you'll be privy to this information. Avoid picking berries near roads, where they might be contaminated by exhaust fumes, or in the vicinity of crop spraying.

Part of the reason for laws barring foraging is that many wild plants are on the brink of extinction. In the United States, these are compiled on the Federal Register of Endangered and Threatened Wildlife and Plants; in addition to this list, there is an interim list of plants that have not yet made it to the federal register but are nonetheless endangered. Generally speaking, edible

The ultimate foraging ground for the serious berry picker is the Scandinavian woodlands, where berries grow so densely underfoot that it is hard for the selective picker to avoid stepping on them.

berries are not significantly represented on these lists, but some of the materials you may gather for ornamental berry arrangements are. States also have individual laws regarding the picking of certain plants. United States residents should obtain the federal register from the government (see "Sources") and contact the state Cooperative Extension Service to learn if there are any berry plants that grow wild in their area that they shouldn't be picking.

In Canada, such matters are handled on a provincial level by offices of Agriculture Canada, Plant Health Division, and Environment Canada (see "Sources"). The Ministry of Natural Resources, located in Ontario, can help you determine what is fair foraging for you. Agriculture Canada maintains an extensive, all-inclusive listing of species endangered worldwide, so rather than obtain a list, it's best to call a local authority and inquire about the berries you want to pick. In Britain, the Native Conservancy Council can provide the latest definitive list of protected plants, and the Ministry of Agriculture (see "Sources") can be consulted on such matters as well.

If you'd rather not rough it, there's also the option of the pick-your-own farms and farmers' markets. In the United States, write to your state agricultural department for the locations of berry farms (see "Sources"). Elsewhere, check with your agricultural department for berry farm locations or consult the *Yellow Pages*. Most pick-your-own farms recommend that you call ahead to make sure weather conditions are good for picking and to find out what's in season.

If you are planning a group berry-picking expedition that will take you into damp, boggy areas, make sure everyone has high, sturdy boots, left. Protective clothing and gloves to shield fingers from thorns are also recommended for the well-outfitted picker.

When harvesting bramble fruits, right, do not pack them too tightly in your container or they will get crushed. Never apply force to remove bramble fruits from their canes. If they do not detach easily, they just aren't ripe enough yet.

If sallying off into the woodlands on a berry hunt seems too rigorous, simply head to a pick-your-own farm, left. In many regions, berry fanciers need only hop out of their cars to have access to fields of straw-berries, raspberries, and blueberries.

233

HARVESTING
THE BERRIES

While there are a number of complicated gadgets for collecting berries—involving bags and scoops—harvesting by hand is quite an efficient method for the more casual berry picker.

In the case of gooseberries, which have sharp thorns, it's wise to wear thick leather gloves. Pick them by raising a branch with one hand and stripping the pendant berries off with the other. If the gooseberries are very ripe, it's best to use them immediately in desserts; if they are underripe, use them in preserves and sauces. Gather the berries in baskets or rigid plastic containers rather than plastic bags (in which berries may get bruised and are likely to sweat). Keep them as cool as possible during the journey home in the car.

Currant picking is also not an easy chore. Removing the currants from the shrub is slow work because each cluster must be pinched off with the thumb and forefinger. Some people avoid back strain by sitting on a low stool while harvesting currants. The longer currants are left on the bush, the sweeter they get. Their pectin content also diminishes with time. If you want to use the currants for jelly, pick them when they are still slightly green. As Alexandre Dumas advised in writings collected in *Dumas on Food*, "It is important in making currant jelly [*gelée de groseilles*] to use currants which are not too ripe and which are still acidulous, so as to keep the jelly really clear."

Always taste blueberries and bilberries before picking them. Sometimes they appear ripe but aren't. You'll know a bush is ripe for picking when the berries easily separate from the stems. However, it's un-likely that all of the fruit in a blueberry cluster will ripen at the same time. To harvest the ripe ones, rake your fingers lightly across the blueberries. The ones that are ready to be eaten will fall right into your bucket. If you carefully watch your blueberries and bilberries, you'll know precisely when they're ripe: blueberries are ready to be picked two to three days after turning entirely blue.

Raspberries separate from their cores when you pick them; other bramble fruits, like blackberries, take the core with them. Raspberries are also very tender. You must handle them gently and not pack too many in a container. You'll know raspberries are ripe when they turn deep red, purple, glossy black, or golden yellow tinged with pink, depending on variety.

Blackberries may go through several color stages in the ripening process, going from red to blue to purple before deepening to black. Of course, picking the thornless kinds is a far more pleasant task then wading through brambles. Blackberries are soft berries—although not as fragile as raspberries—and when ripe they can be pulled very easily off the plant.

When picking strawberries, always be sure to break the stem by pinching it with your thumb and forefinger. Never just pull the berry from the plant, as this can damage the fruit or separate it from its cap—which will cause it to wither.

Always search out the choicest specimens. Among the red varieties, make sure there is no white showing on the berry, as this is considered a defect. Among the white varieties, look for ones that are pristine in color, with no brownish bruise; a slight yellow or pink blush is a normal characteristic of some white strawberries.

If you are accustomed to the traditional, large-fruited strawberry, you may find it a bit time-consuming to pick a substantial quantity of alpine strawberries. Alice B. Toklas experienced this problem and reminisced about it in her *Cook Book*. She and Gertrude Stein tended a garden near the French Alps one summer in which grew *fraises des bois*. One morning Miss Toklas was inspired to go berry picking, but the experience was more stressful than romantic: "It took me an hour to gather a small basket for Gertrude Stein's breakfast, and later when there was a plantation of them in the upper garden our young guests were told that if they cared to eat them they should do the picking themselves."

If the prospect of intensive berry picking dismays you, see if you can enlist children to do the job. Little ones love picking alpine strawberries because searching for the tiny fruits is as much fun as going on a treasure hunt.

California-grown strawberries are harvested well into October. These berries will not be allowed to sit for long and will soon be whisked off to large refrigerators to preserve their freshness.

SOURCES

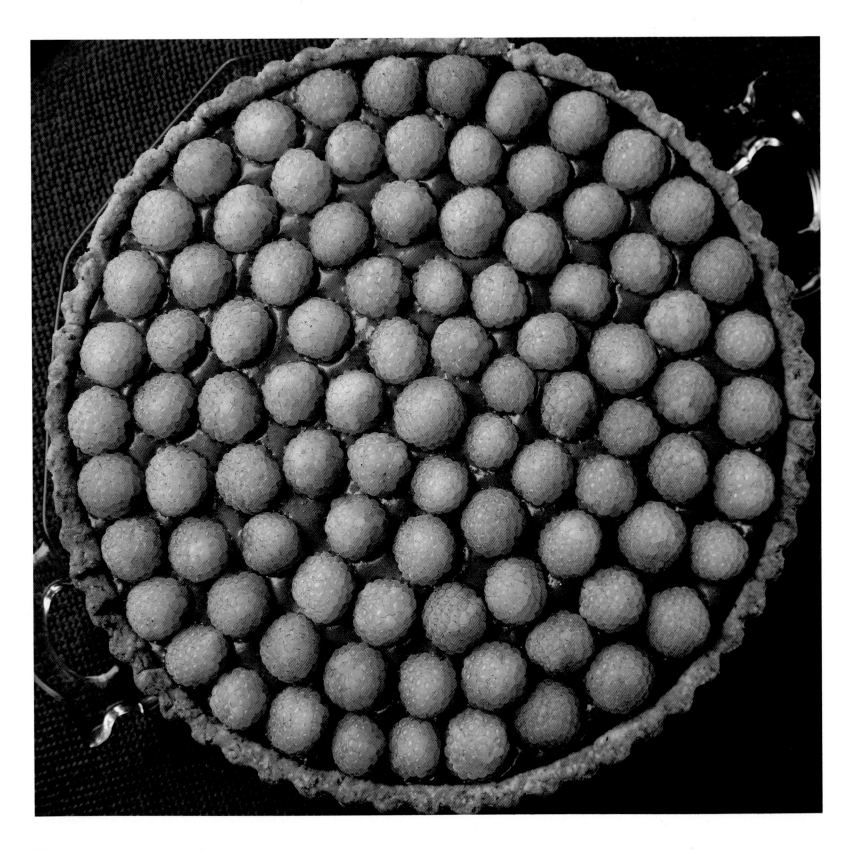

RECIPE LIST

BIBLIOGRAPHY

CULINARY

Beeton, Isabella. *The Book of Household Management*. Jonathan Cape, 1968.

Dumas, Alexandre. *Dumas on Food*. Translated by Alan and Jane Davidson. Oxford University Press, 1982.

Greene, Janet et al. *Putting Food By*. Revised 4th ed. Lexington, Massachusetts: The Stephen Greene Press, 1988.

Grigson, Jane. *Jane Grigson's Fruit Book*. Penguin Books, 1982, 1988.

Johnson, Mireille. *The Cuisine of the Rose*. New York: Random House, 1982.

Norwak, Mary. *The Complete Book of Home Preserving*. Ward Lock, 1978, 1988.

Root, Waverley. *Food: An Authoritative Visual History and Dictionary of the Foods of the World*. New York: Simon and Schuster, 1980.

Simmons, Amelia. *The First American Cookbook. (A Facsimile of "American Cookery," 1796)*. Essay by Mary Tolford Wilson. New York: Dover Publications, 1984.

Toklas, Alice B. *The Alice B. Toklas Cook Book*. Brilliance Books, 1983.

Washington, Martha. *Martha Washington's Booke of Cookery*. Transcribed and annotated by Karen Hess. New York: Columbia University Press, 1981.

FOLKLORE AND HISTORIC USAGES

Bowen, 'Asta. *The Huckleberry Book*. Helena, Montana: American Geographic Publishing, 1988.

Bringle Clarke, Charlotte. *Edible and Useful Plants of California*. Berkeley, California: University of California Press, 1977.

Clark, J.G.D. *Prehistoric Europe: The Economic Basis*. London: Stanford, California: Stanford University Press, 1966.

Fielder, Mildred. *Plant Medicine and Folklore*. New York: Winchester Press, 1975.

Frazer, Sir James George. *The Golden Bough: A Study in Magic and Religion, Part VII, Vol. II*. Macmillan, 1980.

Friend, Hilderic. *Flower Lore*. Rockport, Massachusetts: Para Research, 1981.

Gerard, John. *The Herball or Generall Historie of Plants*. New York: Dover Publications, 1975.

Heiser, Charles B. *Of Plants and People*. Norman, Oklahoma: University of Oklahoma Press, 1985.

Holmberg, Uno. *The Mythology of All Races: Finno-Ugric, Siberian*. New York: Cooper Square Press, 1964.

Kavasch, Barrie. *Native Harvests: Recipes and Botanicals of the American Indian*. New York: Random House, 1979.

Lewis, Naphtali, and Meyer Reinhold. *Roman Civilization, Sourcebook II: The Empire*. New York: Harper & Row, 1966.

Miller, Amy Bess Williams. *Shaker Herbs: A History and a Compendium*. New York: Clarkson N. Potter, 1976.

Thistleton-Dyer, Thomas Firminger. *The Folk-Lore of Plants*. Detroit: Gordon Press, 1968.

Vogel, Virgil J. *American Indian Medicine*. Norman, Oklahoma: University of Oklahoma Press, 1970.

Webber, Ronald. *The Early Horticulturists*. David & Charles, 1968.

Whiteaker, Stafford. *The Compleat Strawberry*. New York: Crown, 1985.

Wigginton, Eliot, ed. *Foxfire 3*. Garden City, New York: Doubleday Books, 1975.

FORAGING AND IDENTIFICATION

Angell, Madeleine. *A Field Guide to Berries and Berrylike Fruits*. New York: Bobbs-Merrill Co., 1981.

Berglund, Berndt, and Clare E. Bolsby. *The Edible Wild*. New York: Charles Scribner's Sons, 1971.

Gibbons, Euell. *Stalking the Wild Asparagus*. New York: David McKay Co., 1962.

Hinds, Harold R., and Wilfred A. Hathway. *Wildflowers of Cape Cod*. Chatham, Massachusetts: The Chatham Press, 1968.

Jordan, Michael. *A Guide to Wild Plants: The Edible and Poisonous Species of the Northern Hemisphere*. London: Millington Books, 1976.

Knutsen, Karl. *Wild Plants You Can Eat: A Guide to Identification and Preparation*. Garden City, New York: Doubleday & Co., 1975.

Lang, David C. *The Complete Book of British Berries*. London: Threshold Books Ltd., 1987.

Peterson, Roger Tory, and Margaret McKenny. *A Field Guide to Wildflowers of Northeastern and North Central North America*. Boston: Houghton Mifflin Co., 1968.

Phillips, Roger. *Wild Food*. Pan Books, 1983.

Rodway, Avril. *Wild Foods*. Brian Trodd, 1988.

GARDENING

Clarke, J. Harold. *Growing Berries and Grapes at Home*. New York: Dover Publications, 1976.

Dennis, John V. *The Wildlife Gardener*. New York: Alfred A. Knopf, 1985.

Dirr, Michael. *Manual of Woody Landscape Plants: Their Identification, Ornamental Characteristics, Culture, Propagation, and Uses*. 3rd ed. Champaign, Illinois: Stikes Publishing Co., 1983.

Fish, Margery. *Gardening in the Shade*. London: Faber and Faber, 1983.

Hahn, Otto. *Macmillan Book of Ornamental Gardening*. Translated by Susan H. Ray. New York: Macmillan, 1985.

Hillier, Harold G. *Manual of Trees and Shrubs*. David & Charles, 1981.

Jekyll, Gertrude. *Gertrude Jekyll on Gardening*. Edited by Penelope Hobhouse. Collins, 1983.

Liberty Hyde Bailey Hortorium. *Hortus Third*. Initially compiled by Liberty Hyde Bailey and Ethel Zoe Bailey. New York: Macmillan, 1976.

Liebster, Gunther. *Macmillan Book of Berry Gardening*. Translated by Carole Ottesen. New York: Macmillan, 1986.

Robinson, William. *The Wild Garden*. Introduction by Robin Lane Fox. London: The Scolar Press, 1977.

Rogers Gessert, Kate. *The Beautiful Food Garden*. Pownal, Vermont: Garden Way Publishing/Storey Communications, 1987.

Sackville-West, Vita. *Vita Sackville-West's Garden Book*. Edited by Philippa Nicolson. Michael Joseph, 1983.

HORTICULTURAL HISTORY

Coats, Alice M. *The Plant Hunters*. (*The Quest for Plants*.) Studio Vista, 1969.

Darwin, Charles. *A Naturalist's Voyage Round the World*. London: John Murray, 1901.

Kellam de Forest, Elizabeth. *The Gardens and Grounds at Mount Vernon: How George Washington Planned and Planted Them*. Mount Vernon, Virginia: The Mount Vernon Ladies' Association of the Union, 1982.

Hedrick, U. P. *A History of Horticulture in America to 1860*. New York: Oxford University Press, 1950.

Huxley, Anthony. *An Illustrated History of Gardening*. New York: Paddington Press Ltd., 1978.

Leighton, Ann. *American Gardens in the Eighteenth Century*. Boston: Houghton Mifflin, 1976.

_____. *Early American Gardens*. Boston: Houghton Mifflin, 1970.

Lyte, Charles. *The Kitchen Garden*. Yeovil, Somerset, U.K.: Oxford University Press, 1987.

McLean, Teresa. *Medieval English Gardens*. Collins, 1981.

Phipps, Frances. *Colonial Kitchens, Their Furnishings and their Gardens*. New York: Hawthorn Books, 1972.

Roach, F. A. *Cultivated Fruits of Britain*. Blackwell, 1985.

Sanecki, Kay N. *Discovering Gardens in Britain*. Aylesbury, Bucks, U.K.: Shire Publications, 1987.

Seebohm, M. E. *The Evolution of the English Farm*. Revised 2nd ed. London: Ruskin House, 1952.

Schlebecker, John T. *Whereby We Thrive: A History of American Farming 1607–1972*. Ames, Iowa: Iowa State University Press, 1975.

SOURCE INFORMATION

(All listings are mail-order companies unless otherwise noted. Call or write for catalogue information.)

FOOD

Elizabeth the Chef
St. Mary's Road
Sydenham Farm Industrial Estate
Leamington Spa CV31 1JP
0926 311531
black currant cakes

Culpepper Ltd.
Hadstock Road
Linton, Cambridge
CB1 6NJ
0223 891196
medicinal berries

J. Floris, Ltd.
89 Jermyn Street
London SW1Y 6YH
01 430 2885
berry potpourris

Fortnum & Mason
Piccadilly
London W1A 1ER
01 734 8040
preserves

Harrods
Knightsbridge
London SW1X 7XL
01 730 1234
fresh berries and berry desserts (sold in store) and preserves

Whittard & Co. Ltd.
111 Fulham Road
London SW3 6RP
01 589 4261
berry teas

North America

PRESERVES AND SPREADS

American Spoon Foods
P.O. Box 566
Petoskey, MI 49770
(616) 347-9030
assorted preserves

Bainbridge's Festive Foods
P.O. Box 587
White Bluff, TN 37187
(615) 383-5157
numerous jellies and preserves, including Fig Berry preserves and Cranberry Pineapple jelly

Boothbay Blues
Joppa Road
P.O. Box 237
West Southport, ME 04576
(207) 633-4677
blueberry, strawberry, and raspberry preserves

Bremen House
220 East 86th Street
New York, NY 10028
(212) 288-5500
Scandinavian berry preserves, including cloudberry and lingonberry; also carries blueberry and elderberry preserves

Cascade Conserves
P.O. Box 8306
Portland, OR 97207
(503) 243-3608
assorted preserves and conserves

Chalif, Inc.
P.O. Box 27220
Wyndmoor, PA 19118
(215) 233-2023
Strawberries & Champagne mayonnaise

Clearbrook Farms
5514 Fair Lane
Fairfax, OH 45227
(513) 271-2053
assorted preserves

The Great Valley Mills
687 Mill Road
Telford, PA 18969
(215) 256-6648
seedless black raspberry, red raspberry, blackberry, blueberry, and strawberry preserves; some no-sugar preserves; whole blueberry syrups and raspberry and strawberry syrups

Green Briar Jam Kitchen
6 Discovery Hill Road
East Sandwich, MA 02537
(508) 888-6870
sun-cooked berry preserves

Hickins
R.F.D. 1
Black Mountain Road
Brattleboro, VT 05301
(802) 254-2146
blackberry, raspberry, and elderberry preserves

House of Webster
Box 488
1013 North Second Street
Rogers, AR 72757
(501) 636-4640
jams and jellies

Knott's Berry Farm
8039 Beach Boulevard
Buena Park, CA 92670
(714) 827-1776

200 Boysenberry Lane
Placentia, CA 92670
(714) 579-2400
full range of preserves

Kozlowski Farms
5566 Gravenstein Highway North
Forestville, CA 95436
(707) 887-2104
(707) 887-1587
blackberry, raspberry, gooseberry,
strawberry jams; no-sugar-added
blueberry conserve; berry syrup-sauces

Maury Island Farming Company
Route 3
Box 238
Vashon, WA 98070
(206) 463-5617
60-percent fruit toppings, jams, and
jellies

Michel's Magnifique
34 North Moore Street
New York, NY 10013
(212) 431-1070
berry relishes to complement
meats and pâtés

Paprikas Weiss
1546 Second Avenue
New York, NY 10028
(212) 288-6117
fresh berries seasonally in store;
berry preserves

Rowena's & Captain Jaap's
758 West 22nd Street
Norfolk, VA 23517
(800) 627-8699
(804) 627-8699
Cranberry Nut Conserve; raspberry-
cranberry preserves

Sarabeth's Kitchen
169 West 78th Street
New York, NY 10024
(212) 580-8335
Apricadabra (apricot, pineapple, and
currant preserves), cranberry relish, Rosy
Cheeks (strawberry and apple preserves)

Wilds of Idaho
1308 West Boone
Spokane, WA 99201
(509) 326-0197
huckleberry jam, dessert toppings, and
dessert filling

VINEGARS, SYRUPS, AND OILS

Bickford Flavors
282 South Main Street
Akron, OH 44308
(800) 283-8322
(216) 762-4666
alcohol-free oils of natural fruits,
including blueberry, blackberry,
raspberry, and strawberry

Chicama Vineyards
Stoney Hill Road
West Tisbury, MA 02575
(508) 693-0309
cranberry, blueberry, and raspberry
vinegars

Country Essences
309 Peckham Road
Watsonville, CA 95076
(408) 722-4549
raspberry and blueberry vinegars

Crabtree & Evelyn
Box 167
Woodstock, CT 06281
(203) 428-2766
vinegars and syrups

Dean & DeLuca
560 Broadway
New York, NY 10012
(800) 221-7714
(212) 431-8369
chutneys, syrups, dried blueberries, berry
vinegars, berries preserved in alcohol

Duggan's Ingredients
1365 Interior Street #A
Eugene, OR 97402
(503) 343-8697
complex oils, vinegars, and dressings
made with whole fresh berries, including
Blackberry Walnut Delight salad dressing
and raspberry and blackberry vinegars

Fraser Morris
931 Madison Avenue
New York, NY 10021
(212) 288-2727
sauces and vinegars

Maison Glass
111 E. 58th Street
New York, NY 10022
(212) 755-3315
vinegars, syrups, and preserves

Matthew's 1812 House Inc.
15 Whitcomb Hill Road
Cornwall Bridge, CT 06754
(203) 672-0149
Cumberland sauce (contains red currant
jelly and is used as accompaniment for
lamb and game)

Paula's California Herb Vinegars and Premium Oils
Sweet Adelaide Enterprises, Inc.
3457-A South La Cienega Boulevard
Los Angeles, CA 90016
(213) 559-6196
Raspberry Royale (white-wine vinegar blended with raspberry syrup, raspberry liqueur, and lavender flower); no mail order, but sells direct to stores and restaurants

S. E. Rykoff
P.O. Box 21467
Los Angeles, CA 90021
(800) 421-9873
imported gooseberries, lingonberry and cloudberry syrups

The Silver Palate
274 Columbus Avenue
New York, NY 10023
(212) 799-6340
vinegars, cranberry and blueberry chutneys, cassis dessert sauce

Whistling Wings Farm, Inc.
427 West St.
Biddeford, ME 04005
(207) 282-1146
gift baskets containing raspberry vinegar, syrup, and honey, as well as jams and jellies

New Zealand and Australia

David Jones Foodhalls
Market Street
Sydney, New South Wales

Bondi Junction, New South Wales

Brookvale, New South Wales

Parramatta, New South Wales

Melbourne, Victoria

Adelaide, South Australia

BEVERAGES

Davidson's Inc.
P.O. Box 11214
Reno, NV 89510
(702) 356-1690
teas, including Wild Strawberry, Spiced Raspberry, and Black Currant in decaffeinated and regular blends

Peter Dent Food and Catering
120 Hudson Street
New York, NY 10013
(212) 219-0666
lingonberry juice, as well as preserves and syrups

A. I. Eppler Ltd.
P.O. Box 16513
Seattle, WA 98116-0513
(206) 932-2211
black currant wine; local sales only

Nashoba Valley Winery
100 Wattaquadoc Hill Road
Bolton, MA 01740
(508) 779-5521
Semisweet and Dry Blueberry wines and Elderberry-Apple Wine; carried throughout New England; call for locations

The Nutmeg Vineyard
800 Bunker Hill Road
P.O. Box 146
Andover, CT 06232
(203) 742-8402
strawberry and raspberry wine; sold on premises and only in Connecticut

Shallon Winery
1598 Duane Street
Astoria, OR 97103
(503) 325-5978
wild evergreen blackberry wine, cranberry-and-whey wine; local sales only

United Society of Shakers
Sabbathday Lake
Poland Spring, ME 04274
(207) 926-4597
strawberry-leaf and raspberry-leaf teas

DESSERTS AND BAKED GOODS

Bickford Flavors
282 South Main Street
Akron, OH 44308
(800) 283-8322
(216) 762-4666
Craisins (dried sweetened cranberries for dipping in chocolate, making chocolate bark, baking in nutbreads, etc.)

Cranberry Sweets Co.
P.O. Box 501
Bandon, OR 97411
(503) 347-9475
preservative-free cranberry candies

Culinary Products
Polly Jean's Pantry
4561 Mission Gorge Place
Suite K
San Diego, CA 92120
(619) 283-5429
dessert sauces, including Blueberry Cassis; numerous preserves and cranberry conserve

Ethel M Chocolates
P.O. Box 98505
Las Vegas, NV 89193
(800) 634-6584
(702) 458-8864
raspberry butter cream chocolates

The Famous Pacific Dessert Company
420 East Denny Way
Seattle, WA 98122
(206) 328-1950
raspberry and marionberry purees

Li-Lac Truffles
120 Christopher Street
New York, NY 10014
(212) 242-7374
raspberry truffles

Miss King's Kitchens, Inc
5302 Texoma Parkway
Sherman, TX 75090
(214) 893-8151
Strawberry Cream Cheese Cake

New England Dairy Foods
398-400 Pine Street
Burlington, VT 05401
(800) 447-1205
(802) 864-7271
Strawberry Swirl and Blueberry Swirl
cheesecakes

Pan Handler Products
4580 Maple Street
Waterbury Center, VT 05677
(802) 253-8683
dessert sauces, including Cranapple
Maple, Apple Blueberry, Strawberry
Amaretto, and Raspberry Apple

Wolferman's
1 Muffin Lane
P.O. Box 15913
Lenexa, KS 66215-5913
(800) 255-0169
blueberry muffins and preserves

DECORATIONS AND MISCELLANEOUS USAGES

Aphrodisia
282 Bleecker Street
New York, NY 10014
(212) 989-6440
medicinal dried berry mixtures and berry-
leaf teas

Faith Mountain Herb
Main Street
P.O. Box 199
Sperryville, VA 22740
(703) 987-8824
berry potpourris, rosehips, berry wreaths

The Fragrance Shop
49 Old Concord Road
Henniker, NH 03242
(603) 746-4431
rosehips by the cupful, berry potpourris
(mid-April through Thanksgiving)

The Golden Pineapple
49 Old Concord Road
Henniker, NH 03242
(603) 746-4431
berry wreaths (Thanksgiving through mid-
April)

The Herbfarm
32804 Issaquah—Fall City Road
Fall City, WA 98024
(206) 784-2222
medicinal berries (rosehips, juniper
berries), holiday wreaths and garlands
with berries, berry leaf teas

Rathdowney Herbs and Herb Crafts
3 River Street
Bethel, VT 05032
(800) 543-8885
(802) 234-9928
berry potpourris and wreaths, dried
medicinal elderberries and juniper
berries

The Rosemary House
120 South Market Street
Mechanicsburg, PA 17055
(717) 697-5111
rosehip and sumac berry wreaths, berry
dyes, berry teas

ORGANISATIONS

Ministry of Agriculture, Fisheries
and Food
Whitehall Place London SW1A 2HH
01-270-3000
contact regarding endangered plant
species

Fruit Group
Royal Horticultural Society
Attn: Mrs. M. Sweetingham
Vincent Square, London SW1P 2PE
01 834 4333

Mr. Harry Baker
Fruit Officer
Wisley Gardens
Woking
Surrey GU23 6QB
0483 224234

National Council for the Conservation of
Plants and Gardens
The NCPG hold National Reference
Collections of some genera of berry
plants.

Gooseberry; *Sorbus* spp and cvrs
Dr. R.A. Benton
Department of Environmental Biology
University of Manchester
Williamson Building
Oxford Road
Manchester M13 9PL

Ilex
Mr J.D. Bond
Crown Estate Office
The Great Park
Windsor
Berkshire SL4 2HT

Pyracantha
Mr M. Stimson
Writtle Agricultural College
Chelmsford
Essex CM1 3RR

Ribes spp and primary hybrids
Mr P. Orris
University Botanic Gardens
Cambridge CB2 1JF

Rubus
Mr B. Gilliland
Clinterty Agriculture College
Kinellar
Aberdeen AB5 0TN

Sambucus
Mr G. Moon
National Trust
Wallington Hall
Cambo
Morpeth
Northumberland NE61 4AR

United States
(Some organizations are international.)

Appalachian Trail Conference
P.O. Box 807
Harpers Ferry, WV 24525-0807
(304) 556-6331
contact for listings of Appalachian Trail
guides

Blueberry and Cranberry Research Center
Rutgers University
Chatsworth, NJ 08019
(609) 726-1590

Driscoll Strawberry Association Inc.
P.O. Box 111
Watsonville, CA 95077
(408) 726-3531

International Ribes Association
18200 Mountain View Road
Boonville, CA 95415
worldwide group open to the home
gardener and botanist alike

North American Blueberry Council
P.O. Box 166
Marmora, NJ 08223
(609) 399-1559

North American Fruit Explorers
c/o Jill Vorbeck
Route 1, Box 94
Chapin, IL 62628
quarterly journal *Pomona* reports on
findings of berry test groups

Canada

British Columbia Blueberry Assn.
31852 Marshall Road
Abbotsford, BC V254N5
(604) 859-4200

New Brunswick Strawberry Growers
Association
1115 Regent Street
Fredericton, NB E3B 3Z2
(506) 452-8101

Australia

New South Wales Strawberry Growers
Association
(02) 626-9108

New South Wales Central Tablelands
Berry Fruit Growers Association
(063) 628 851

Victorian Research Station
Horticultural Research Institute
Noxfield, Victoria
(03) 221-2233

Australian Blueberry Growers Association
Victoria
(056) 294 258

Australian Berry Fruit Growers
Association
New South Wales
(063) 628 851

New Zealand

New Zealand Berryfruit Growers
Federation
P.O. Box 10050 Wellington, N.Z.
I.C.L. House
126 The Terrace
735 387

New Zealand Berryfruit Growing
Federation
Berryfruit Market and Licensing Authority
P.O. Box 10050
Wellington
(04) 735 387

(Call ahead for hours.)

(NT = The National Trust)

Barnsley House
Nr. Cirencester
Gloucestershire
028574 281
kitchen garden with berries

Barrington Court (NT)
Nr. Ilminster
Somerset TA19 0NQ
0460 41480
walled kitchen garden with berries;
nursery

Bodnant Gardens (NT)
Tal-y-Cafn
Colwyn Bay
Clwyd LL28 5RE
0492 650460
ornamental berries

Dunham Massey (NT)
Altrincham
Cheshire WA14 4SJ
061 941 1025
collection of *Skimmia*

East Lambrook Manor Garden
South Petherton
Somerset
0460 40328
ornamental berry plantings and nursery

Felbrigg Hall (NT)
Norwich
Norfolk NR11 8PR
026 375 444
kitchen garden with berries

Forde Abbey Gardens
Nr. Chard
Dorset
0460 20231
walled kitchen garden containing
ornamental berries and nursery

Greys Court (NT)
Rotherfield Greys
Henley-on-Thames
Oxfordshire RG9 4PG
049 17 529
kitchen garden with berries

Gunby Hall (NT)
Gunby
near Spilsby
Lincolnshire PE23 5SS
kitchen garden with berries

Hare Hill (NT)
near Macclesfield
Cheshire SK10 4QB
collection of *Ilex*

Knightshayes Court (NT)
Bolham
Tiverton
Devon EX16 7RQ
0884 254665
ornamental berries

Little Moreton Hall (NT)
Congleton
Cheshire CW12 4SD
0260 272018
standard gooseberries

The Royal Horticultural Society's Garden,
Wisley
Woking
Surrey GU23 6QB
0483 224234
model backyard berry gardens

Tintinhull House (NT)
Tintinhull
Nr. Yeovil
Somerset BA22 9PZ
0935 822509
kitchen garden with berries

Upton House
Banbury
Oxfordshire OX15 6HT
029587 266
walled kitchen garden with extensive
berry plantings

Waterperry Horticultural Centre
Ickford
Nr. Oxford
08447 254
comprehensive berry fields

United States

Ashumet Holly Reservation and Wildlife
Sanctuary
Ashumet Road
East Falmouth, MA 02536
(508) 563-6390

Cape Cod National Seashore
Province Lands Visitor Center
Race Point Road
Provincetown, MA
for information contact:
Cape National Seashore
South Wellfleet, MA 02663
(508) 487-1256
wild cranberry bogs (mid-April to late
October)

Cornell Plantations
Ronald U. Pounder Heritage Garden
Heritage Plantations
One Plantations Road
Ithaca, NY 14850-2799
(607) 255-3020
re-creation of nineteenth-century kitchen
garden, with berries

Cranberry World
Ocean Spray Cranberries Inc.
Plymouth, MA 02360
(508) 747-1000
(508) 747-2350
historic exhibits; bog tours in October

Longwood Gardens
U.S. 1 and Pa. 52
Kennett Square, PA 19348-0501
(215) 388-6741
berry garden

Mount Vernon
Mount Vernon Ladies' Association
of the Union
Mount Vernon, VA 22121
(703) 780-2000
re-creation of George Washington's
kitchen garden, with extensive berry
plantings

Van Cortlandt Manor
Route 9
Croton-on-Hudson, NY
for information contact:
Historic Hudson Valley
150 White Plains Road
Tarrytown, NY 10591
(914) 631-8200
old-fashioned kitchen garden with a
variety of berries; holiday-season berry
decorations

New Zealand

Shaws Berry Farm
Rangitopuni Road
Albany, North Auckland
R.D. 3, Albany
(09) 412 8953

Willow Park Group
Addington Road, Otaki
(069) 47293

NURSERIES

PLANTS

Burston-Tyler Rose & Garden Centre
North Orbital Road
Chiswell Green
St. Albans, Herts. AL2 2DS
0727 324 44
extensive rose selection

Clifton Nurseries Ltd.
5A Clifton Villas
London W9 2PH
01 289 6851
holly

C. W. Groves & Son
West Bay Road
Bridport, Dorset DT6 4BA
0308 336 54
strawberry specialists

Kingsley Strawberries Ltd.
Headley Mill Farm
Standard Lane
Headley, Bordon, Hants. GU35 8RH
042 03 2322
extensive alpine strawberry stock

Ken Muir
Honeypot Farm
Rectory Road
Weeley Heath, Clacton-on-Sea, Essex
CO16 9BJ
0255 830 181
full range of edible berries

James Trehand & Son Ltd.
Camellia Nursery
Stapehill Road
Hampreston, Wimborne, Dorset
BH21 7NE
0202 873 490
blueberry plants

Wyvalley Landscape & Maintenance Co.
Cow Lane, Laverstock,
Salisbury, Wiltshire SP1 2SR
0722 201 29
cotoneasters and other ornamental shrubs
and trees

BERRY CAGES

Agriframes Ltd.
Charlwoods Road
East Grinstead, Sussex RH19 2HG
0342 286 44

Taurus Products Ltd.
Unit 5, Lexden Industrial Estate
Crowborough, East Sussex TN6 2EG
089 26 5369

United States

Ahrens Strawberry Nursery
R.R. 1
Huntingsburg, IN 47542
(812) 683-3055
strawberries and other berries

Bluestone Perennials
7211 Middle Ridge Road
Madison, OH 44057
(216) 428-1327
ornamentals

Brittingham Plant Farms
Box 2538
Salisbury, MD 21801
(301) 749-5153
strawberries, brambles, and blueberries

Dutch Gardens
P.O. Box 200
Adelphia, NJ 07710
(201) 780-2713
ornamentals

A. I. Eppler Ltd.
P.O. Box 16513
Seattle, WA 98116-0513
(206) 932-2211
44 varieties of black currants, 9 varieties
of white currants, and 10 varieties of red
currants (all European varieties); also
carries jostaberries, gooseberries (green,
yellow, purple, red, and black)

Harris Seeds
Moreton Farm
3670 Buffalo Road
Rochester, NY 14624
(716) 594-9411
alpine strawberry seeds

Makielski Berry Farm and Nursery
7130 Platt Road
Ypsilanti, MI 48197
(313) 572-0060
specializes in raspberries and currants;
also gooseberries, blueberries,
blackberries, and strawberries

Owens Vineyard and Nursery
Route 1, Box 25
Gay, GA 30218
(404) 538-6938
rabbiteye blueberries

Rayner Bros.
Box 1617
Salisbury, MD 21801
(301) 742-1594
strawberries, brambles, blueberries

Roses of Yesterday and Today
802 Brown Valley Road
Watsonville, CA 95076
(408) 724-3537

White Flower Farm
Route 63
Litchfield, CT 06759-0050
(203) 496-9600
ornamentals and alpine strawberry plants

INDEX

254

CREDITS

DECORATIONS

FOOD AND GIFTS

GROWERS